Population Dynamics of
a Philippine Rain Forest People

Population Dynamics of a

University Press of Florida
Gainesville Tallahassee Tampa Boca Raton
Pensacola Orlando Miami Jacksonville

John D. Early
Thomas N. Headland

Philippine Rain Forest People

The San Ildefonso Agta

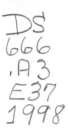
Copyright 1998 by the Board of Regents of the State of Florida
Printed in the United States of America on acid-free paper
All rights reserved

03 02 01 00 99 98   6 5 4 3 2 1

All photographs copyrighted as listed in captions. Photo at beginning of chapter 1
by Janet Headland, 1963; remaining chapter-opening photos by John Early, 1994.
Frontispiece. Right: A traditional Agta hunter shooting an arrow. (Thomas Headland, 1967.)
Left: An Agta boy on a bulldozer. (John Early, 1994.)

Library of Congress Cataloging-in-Publication Data
Early, John D.
Population dynamics of a Philippine rain forest people: the San Ildefonso Agta / John D. Early,
Thomas N. Headland.
p. cm.
Includes bibliographical references and index.
ISBN 0-8130-1555-3 (cloth: alk. paper)
1. Agta (Philippine people)—Population. 2. Agta (Philippine people)—Social conditions.
3. Agta (Philippine people)—Economic conditions. 4. Logging—Philippines—San Ildefonso.
5. Rain forest ecology—Philippines—San Ildefonso. 6. San Ildefonso (Philippines)—Population.
7. San Ildefonso (Philippines)—Social conditions. 8. San Ildefonso (Philippines)—Economic
conditions. I. Headland, Thomas N. II. Title.
DS666.A3E37 1998
304.6'2'089991105991—dc21   97-41150

The University Press of Florida is the scholarly publishing agency for the State University
System of Florida, comprised of Florida A & M University, Florida Atlantic University,
Florida International University, Florida State University, University of Central Florida,
University of Florida, University of North Florida, University of South Florida, and University
of West Florida.

University Press of Florida
15 Northwest 15th Street
Gainesville, FL 32611
http://nersp.nerdc.ufl.edu/~upf

Dedicated to the memory of
Father Morice Vanoverbergh (1885–1982),
missionary, linguist, ethnographer, Philippinist,
who took the first Agta survey in 1936
T.N.H.

To Stefanie, Robby, David, Craig, and Kristine,
who are Grandpa's reminders of the
demographics of the continuing life cycle
J.D.E.

Contents

Tables

Figures

Maps

Preface

The purpose of this book is an investigation of the population dynamics of an Agta Negrito people living on Luzon Island in the Philippines. Population dynamics are the patterns of increase or decrease of a population resulting from its fertility, mortality, and migration. The present study is exceptional in that it presents the results of a 44-year quantitative database of these demographic variables from the time when the Agta were foragers in the rain forest to the period when they became peasants within rural Philippine society. This book is one of only two studies that have completely reconstructed the population dynamics of a foraging group without relying on mathematical models. The assumptions of those models may be of questionable validity for this type of population. As background for understanding the population dynamics, the preliminary chapters outline pertinent points of Agta ethnography and Philippine history as compiled from the existing literature.

This book may be seen as a case history of the social and population dynamics taking place in frontier regions remote from the media centers of the international world. It describes an example of a minority people found in such regions who are without economic, political, and social power and who are so little known that their story is not well understood. The analysis herein shows the impact of international logging interests on tropical rain forests as well as the condition of landless peasantry seeking a way of survival. This study also describes diseases suffered by these groups, which may help in confronting the medical problems of traditional indigenous peoples.

The subject matter of the book is within the disciplines of cultural and demographic anthropology and as such is primarily addressed to demographers and anthropologists. Some demographers and anthropologists may find the treatment overwritten in that explanations are given of terms and procedures for which a professional in these fields would need no explanation. We ask their indulgence of the effort to make the work intelligible to a wider audience, especially those concerned with population, ecological, and health issues.

The coauthors are indebted to a number of people without whose assistance this research could not have been accomplished. Our wives were coworkers. Janet Headland spent innumerable hours coding data based on her intimate knowledge of the Agta people; her many other roles will be referred to in the following chapters whenever the Headlands are mentioned. Jacky Early spent many hours entering data into the computer and performing numerous backup tasks. Professors Michael Billig of Franklin and Marshall College, Bion Griffin of the University of Hawaii, Barry Hewlett of Washington State University, and Rachel Headland Ulmer read preliminary drafts of the manuscript and offered valuable suggestions. Tomas Casala of Casiguran, Professor James Eder of Arizona State University, and Professor Kim Hill of the University of New Mexico furnished unpublished data on the Agta, Batak Negritos, and Ache, respectively. Professor Bill Hunt of Florida Atlantic University wrote some of the computer programs. The New Tribes missionaries, Iris Harrison and Anne Kueffer Quirk, generously kept records of Agta births and deaths from 1980 to 1982 while the Headlands were in the United States. Mrs. Away Aduanan, Mr. Eleden Aduanan, Mrs. Maming Alamra, and Mrs. Pompoék Saguné́d are four Agta elders with expert knowledge of genealogical histories who worked many hours with the Headlands in 1992 and 1994. Eleden Aduanan's picture is the introductory photograph to chapter 3.

The Summer Institute of Linguistics provided support for the Headlands during their years of residence in the Philippines. Acknowledgment and gratitude are expressed to the Pew Charitable Trusts and the LSB Leakey Foundation for grants which financed follow-up visits to the Philippines for the critical process of correcting and completing the database. Gratitude is also expressed to Gelhardt Graphics of Tallahassee, Florida, for digitizing and editing the photographs.

This book is dedicated, in part, to the memory of Morice Vanoverbergh. Father Vanoverbergh, CICM, went to the Philippines from Belgium in 1909 as a Roman Catholic missionary priest and died there in 1982. There are no Philippine scholars who have not benefited from Father Vanoverbergh's publications on Philippine cultures and languages. The present volume is no exception. He made a research trip to the Casiguran Agta in 1936 and in less than half a year there collected an amazing amount of information on their culture and language. These data included, to our good fortune, a list of 476 Agta names (Vanoverbergh 1937–38:139–47), which proved invaluable for corroborating our own Agta informants' memories. Biographical details of Vanoverbergh's life and academic accomplishments may be found in *Anthropos* 78 (1983, pp. 872–73), *Philippine Quarterly of Culture and Society* 3 (1975, pp. 201–3), and *Saint Louis University Research Journal* 13 (1982, pp. 423–83).

Part I

Introduction

Chapter 1

The San Ildefonso Agta

This research investigates the population dynamics of a Negrito group in the Philippines who call themselves Agta. Population dynamics describe the size of a population, its age-sex composition, and changes over time owing to the interplay of fertility, mortality, and migration. These demographic variables, in turn, are influenced by cultural and biological factors. The Agta historically lived by hunting and gathering in the rain forests of eastern Luzon. This and the next two chapters sketch pertinent points from the history and ethnography of the Agta as background for understanding the population dynamics of the San Ildefonso Agta. (See the note at the end of this chapter on the ambiguity and usage of the terms *Agta, Filipino,* and *lowlander.*)

THE PEOPLE AND THE PLACE

The Negritos

Tiny groups of Negritos are scattered throughout the forests of the Philippines and other countries of Southeast Asia. Physically these people are distinguished by dark skin color, fuzzy hair, and short stature. Men average 152 centimeters (5 feet) in height and women 141 centimeters (4.5 feet). They are frequently referred to as pygmies, a term more correctly used for the central African Pygmy populations. They have traditionally been known as aboriginal inhabitants of the rain forest and use bows and arrows to hunt large game. Most of Southeast Asia's Negrito populations are rapidly disappearing, far fewer in number in the 1990s than they were 100 to 300 years ago. Several Negrito groups in the Andaman Islands have become extinct in the last 100 years. The four remaining Andaman Negrito groups have declined greatly since the first census there in 1901, with the Onge, for example, dwindling from 1,000 then to 96 in 1988. The 10 Negrito dialect groups in peninsular Malaysia number only 1,800 today, far fewer than in the last century. The Negrito groups in Thailand have declined to only 300 people.

Most of Asia's Negritos are found in the Philippines, where 29 ethnolinguistic populations live on six of the major islands in that country. These Negritos have also greatly diminished in number since 1600, from around 10 percent of the Philippine population then, which was less than a million, to only 0.05 percent of that nation's population today. In 1994 these 29 populations numbered approximately 31,000 (Griffin and Headland 1994:71). These groups refer to themselves by terms such as Aeta, Agta, Alta, Arta, Ata, Ati, Atta, Batak, and Mamanwa. Outsiders usually refer to them as Negritos.

Luzon has the largest number of Negritos, who reside in the mountains of Zambales, Bataan, Western Pampanga, Western Tarlac, Southwestern Pangasinan, and in the Sierra Madre range, which rims the eastern side of Luzon. The Negritos of the Sierra Madre refer to themselves and their language by the term Agta. Filipinos usually call them "Dumagats." They number about 9,000 and are divided into 10 ethnolinguistic groups (Griffin and Headland 1994:72). Some Agta distinguish two types of Agta groups (Estioko-Griffin and Griffin 1975:237; Rai 1990:67–68). The first live in the mountainous areas at some distance from both the coast and the towns. They rely more on hunting and gathering and have less contact with Filipino lowland farmers. The second type live closer to farming settlements and interact on a daily basis with a wider range of lowlanders. They are more involved in casual farm labor in addition to hunting, fishing, and gathering. The latter group frequently refer to the former group by the derogatory term *ebuked,* which can be roughly translated as "hillbilly."

The subjects of this study are a small group of Agta Negritos living on the San Ildefonso Peninsula. The research examines their population dynamics for the 44-year period from 1950 to 1993, during which time the group has undergone considerable change. This chapter focuses on the regional geography and the culture of these Agta as it was until the 1960s. Chapter 2 discusses how changes at the national and international levels have affected the peninsula and the wider Casiguran region. Chapter 3 examines how these changes have affected the Agta of this region beginning around 1960. With these chapters presenting the background, chapter 4 begins the demographic examination.

The Geographical Setting

The San Ildefonso Peninsula is located in northeastern Luzon (see Map 1), in the northern part of Aurora Province, in the municipality of Casiguran (see Map 2). To the north of Aurora lies the province of Isabela, to the south is Quezon Province, and to the west are the provinces of Quirino, Nueva Vizcaya, and Nueva Ecija. The coordinates of Aurora Province are: east longitudes 121° 31′ 2″ and 122° 1′ 30″; north latitudes 15° 31′ 43″ and 16° 31′ 00″. The town of Baler in the southern part of Aurora is the provincial capital and is about 230 kilometers (143 miles) from Manila.

Administratively Aurora Province is divided into eight municipalities, each of which has a town center. Historically the majority of the small number of Filipinos in the province resided in these towns but had agricultural land outside them. The municipality of Casiguran has an area of 566.68 square kilometers (219 square miles), which is 18 percent of the land area of Aurora Province.

The town of Casiguran is the center for the municipality of the same name. It is located on the Casiguran River, a short distance inland from Casiguran Bay (Map 1). Southeast of the town lies the San Ildefonso Peninsula. The peninsula is approximately 32 kilometers (20 miles) long and ranges from 2.4 to 7.2 kilometers (1.5 to 4.5 miles) in width. It covers an area of 108.79 square kilometers (42 square miles). Casiguran Bay separates most of the peninsula from the narrow coastal strip and the Sierra Madre Mountains. The peninsula is mountainous, with Mount Disigisaw the tallest peak at 653 meters (2,144 feet). The hilly terrain runs in a northeast-southwest direction, with the higher land tending to run along the eastern side of the peninsula and in places forming cliffs on the eastern coast facing the Pacific Ocean. Until the twentieth century the peninsula was inhabited only by Agta.

The Remoteness of Northeastern Luzon

For the purposes of this study, it is important to understand the geographical inaccessibility of northeastern Luzon from the rest of Luzon and the Philippines as well as the concomitant difficulties of transportation. The remoteness is created by the mountains, the ocean, and the inclement

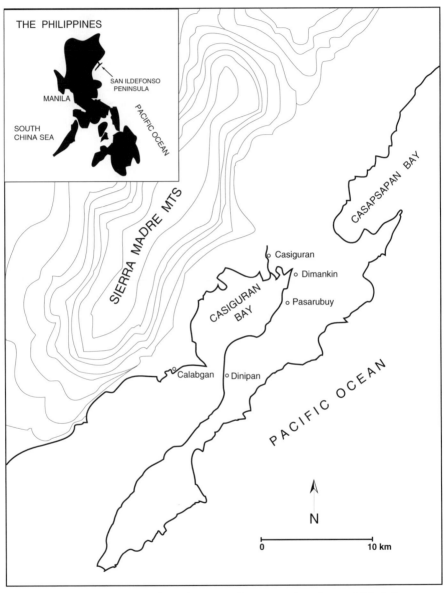

Map 1. San Ildefonso Peninsula and Surrounding Area (based on PCGS, Solano, 1986).

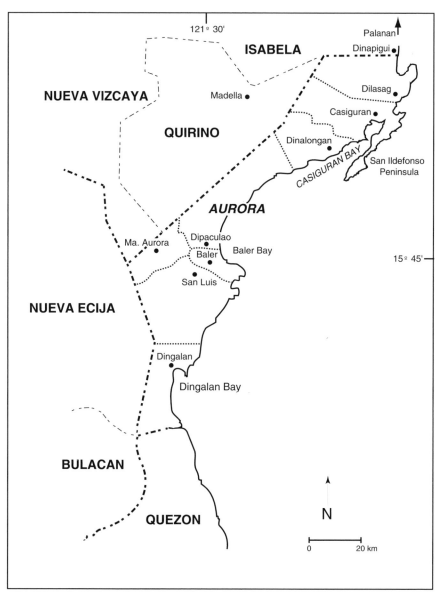

Map 2. Aurora Province and Its Municipalities (based on National Mapping and Resource Information Authority, 1991:3).

weather. Aurora Province is a long, narrow strip of land running in a north-east-southwest direction (Map 2). It is roughly defined on the west by the ridge of the Sierra Madre Mountains, which rise to a height of approximately 1,500 meters (5,000 feet). The eastern slopes of these mountains descend precipitously to a very narrow coastal strip or directly into the Pacific Ocean.

In the Casiguran area the Sierra Madre range is up to 32 kilometers (20 miles) wide when measured in a straight line. However, the precipitous slopes are broken with many crevices and narrow gorges. The only means of traversing the Sierra Madre range between Casiguran and the Cagayan Valley of north-central Luzon is by rough foot trails departing eastward from the town of Madella in Quirino Province. This is a rugged and exhausting journey even for the quick-moving Agta. The trip can take several days when the group includes women and children or if they are carrying cargo. There have been a few logging roads in the past that formed a temporary transportation link, but none of them has survived the force of rainwater turned into cascading streams.

Within the Casiguran municipality, 28 percent of the land area consists of mountains with more than 50 percent slope and 32 percent with 30 percent to 50 percent slope. Only 15.3 percent of the land is flat to gently undulating (less than 8 percent slope) and 4.4 percent undulating to rolling (slope of 8 percent to less than 18 percent) (NAMRIA 1991:7). Because of the steep slopes, agricultural land is scarce. A United Nations report on the Philippines (United Nations 1978:289n.3) notes that only land with slopes from 0 percent to 12 percent can currently be cultivated. Land with slopes from 12 to 25 percent can be safely cultivated only if adequate and special soil management programs are practiced, but these are beyond the present level of the area's agricultural technology.

Movement within the area is also hindered by heavy rainfall evenly distributed throughout the year. The average annual rainfall is 3393 mm (134 inches), with 203 rainy days per year or about 17 days per month. Heat and humidity add to the difficulties of movement. The mean average temperature is 26.45 degrees Celsius per year (79.6 degrees Fahrenheit), with a range of 28.2 degrees to 23.9 degrees Celsius (NAMRIA 1991:15; see Headland 1986:451–52). The average annual humidity is 85 percent, with January the most humid month and May to July the least humid.

Historically the main means of entry into the Casiguran area for people or cargo was from Baler, by foot or in small coastal outriggers. This is a distance of more than 84 kilometers (52 miles). Travel by foot was hazardous because until 1977 there existed only a narrow footpath that alternated between the coastal strip and the lower mountain flanks, where there was no coast. The frequent rainfall increased the hazard. Travel by outriggers hugging the coast was dangerous because of the rough seas and the northeast monsoon winds that threatened to overturn the boats, even the motorized outriggers used after World War II.

Northeastern Luzon also includes the mountains and narrow coastal strips of the provinces of Cagayan and Isabela, which are north of Aurora. Historically the isolation of this region has offered refuge to armed dissenters—such as General Emilio Aguinaldo fleeing from the Americans in 1902 and later, in 1971 and 1972, to Commander Dante, the founder of the antigovernment New People's Army (NPA), during the Ferdinand Marcos regime. A small number of lowlanders and the Agta coexisted in this regional remoteness in dependence on each other. However, the isolation was not complete. Lowlander families would occasionally travel to Madella or Baler or as far as Manila to buy and sell goods as well as visit with friends and relatives. At times the area would be visited by traders seeking forest resources, but neither the Agta nor the lowlanders were economically dependent on such activity.

THE FORAGING AGTA

Traditional Agta Camps and Subsistence

Historically the Casiguran Agta lived away from the towns and preferred areas with easy access to the ocean or mountainous river areas, where game abounded. The peninsula was an ideal living area because of its wide variety of biotopes and ecotones, each with different resources: dense, full-growth forest teeming with deer, wild pig, and monkey; swidden plots; secondary forests; open meadows for driving game with fire; coral reefs; open sandy beaches; mangrove swamps; a sheltered leeward coast to live on during the northeast monsoons; and an alternate sheltered eastern coast when the southwest monsoons blew. All of these alternatives were within less than a day's walk of any Agta camp on the peninsula. There were other benefits to living on the peninsula. The wide bay provided protection against Agta raiding parties from the western side of the Sierra Madre, who at times harassed those living on the coastal strip adjacent to the bay. Finally the peninsula was close enough to the town of Casiguran so that the Agta could make weekly trips there to trade for rice, tobacco, nipa palm wine, and other items.

The Agta lived in residence areas frequently referred to as *camps* by observers. The typical characteristics of these camps were their physical simplicity, temporary nature, and the bonds of kinship among the occupants. In the forest the camps were usually located in a clearing or dry streambed. They were seldom located under the forest canopy to avoid mosquitoes, leeches, dampness, and the danger of falling limbs during rain squalls. The typical shelter was a thatch-covered frame made of poles forming either a one- or two-sided lean-to or a small rectangular dwelling with a split palm or bamboo floor a foot or two off the ground. Headland (1986:460) measured 129 of these shelters and found the average floor area was 3.9 square meters (4.7 sq. yd.) with a per capita floor space of 1.2 square meters (1.4 sq. yd.). Camps were also located on the beaches when the Agta were seeking marine resources (as shown in Figure 1.1). The Agta spend most of their time outdoors, with their dwellings serving primarily as storage for their few possessions, sleeping and cooking areas, and refuge from rain and night dew.

Agta families are nuclear in structure and usually form a household. Camps average six (with a range from three to seven) related nuclear households (Headland 1987b:265). There is no social structure beyond kinship binding together Agta camp groups. However, there is great flexibility in Agta social groupings. The location of the camps as well as their composition frequently changes. Agta families move often for numerous economic and social reasons. A camp group will frequently split up when moves take place. Rai (1990:58) reports that the Disabungan Agta in Isabela Province moved their camp 20 times in 1980, an average of every 18 days. Clark (1990:13) found an average of 10 moves per year by two Agta bands in Cagayan Province. The Agta are a mobile people, but unlike pastoral nomads their moves are not seasonal, continual, or predictable. Every Agta has a "home" river valley where his or her family has historically lived. Most moves are within the river valleys of the family or that of the spouse.

Agta families make their own decisions about their daily activities. Traditionally the Agta have daily set out from their camps to hunt (see Figures 1.2 and 1.3) or forage in the forests or fish the streams and off-shore waters. Oral histories testify to the abundance of game found in the Casiguran area before World War II. An old Agta woman said her father could get three wild pigs in an hour. A lowlander said that in 1917 her family could trade one leaf of tobacco to the Agta for a quarter of a pig or deer and sometimes for a whole animal. The Agta also conducted game drives on the peninsula. The men would lie in ambush at the top of a ridge just inside

Figure 1.1. Typical Agta campsite on the beach. (T. Headland, 1962.)

the forest. The women would string themselves out in a line along the beach at the bottom of a steep hillside meadow and fire the grass. As the fire swept up the hill, it drove the game into the line of waiting men. In the 1950s the Agta could secure as many as 20 pigs or deer in a single drive. In the 1960s, when Thomas and Janet Headland accompanied them on these drives, the Agta would get only one or two. At other times they would head to the beaches for fishing or gathering other types of marine life. The Agta have also practiced occasional swidden (slash-and-burn) horticulture in addition to hunting, fishing, and foraging for plants. In the camps men and women make the various tools used in these activities and prepare the food for consumption (see Figures 1.4 and 1.5). (More detailed accounts of Agta camps, foraging, and horticulture are contained in Griffin 1984a; Headland 1986, 1987b; and Rai 1990. For a complete bibliography on Agta, see Headland and Griffin 1997.)

The Logic of Nonmarket Exchange

For centuries Agta families exchanged goods and services among themselves and also with lowlanders in the area. Since changes in the logic of these exchanges will become important, and since exchanges not based on market logic are frequently misunderstood, some explanation may be helpful. Typically in nonmarket societies, individuals have rights to receive and

Figure 1.2. An Agta hunter with bow and arrows. (T. Headland, 1967.)

Figure 1.3. Agta hunters return to camp with deer meat. (J. Headland, 1975.)

Figure 1.4. Agta hunter singes the hair off a monkey before butchering it. (T. Headland, 1966.)

Figure 1.5. Agta man builds a fish trap. In the background is the frame of an unfinished house made of poles. Palm thatch needs to be added for the roof and those walls that will be covered. (T. Headland, 1966.)

obligations to give goods and services to each other. These rights and obligations are defined by kin relationships between individuals and are called *reciprocity systems* in anthropological literature. Wiessner's (1982:64–65) insurance analogy of pooled risk is useful. The logic of this type of exchange system can be expressed as: from the kin who has to the kin who is in need. It is assumed that both parties will at different times have surpluses or be in need. Where the relationship between the parties is egalitarian, the exchanges are seen as balancing out over time, although no strict accounting is kept. These kinship-exchange relationships have a strong moral or ethical quality and usually cannot be ignored without group sanction, including ostracism. For individuals socialized in competitive market economies with their logic of supply and demand, reciprocity systems may sound utopian. But Peterson (1993) has shown how such systems can also be used in an exploitative manner.

Exchange Relations among Agta

Agta exchange goods and services among themselves based primarily on kin relationships, but residence and individual histories also play a role. Campsite relatives and close affinal relatives are of special importance, as shown in Figure 1.6. This case history from the Headlands' journal illustrates the sharing pattern:

> On October 18 [1983] Dokdoék and her married son went to visit Kiyakin [the husband of the daughter of the sister of Dokdoék's husband]. He was harvesting his swidden rice. Several weeks earlier, Kiyakin had invited Dokdoék to come during the harvest so he could give her some rice. She and her son arrived while the rice was being cut. They joined in and harvested for an hour. When they said they were departing, Kiyakin handed them a sack and told them to take some rice home to his "uncle" [Dokdoék's husband]. Dokdoék said, "Oh, no, we would be ashamed to take your rice. We have plenty at home." Kiyakin insisted. She and her son then stuffed the sack full of unthreshed rice, much more than they had harvested. After it was threshed and dried, the unhusked grain measured 48.6 liters. This was worth the equivalent of six days' wages for agricultural laborers.

The sharing seen in this example is an oft-repeated pattern among the Agta. Two weeks later the same woman received 72 liters of unhusked rice, the equivalent of 10 days' wages, from the family of the daughter of her husband's brother. In another case the Headlands documented, two Agta families (one of a father and the other of his married son) planted 0.54 hectares in rice. Throughout that growing season, only those two families lived at their swidden site. But at harvest time, they were joined by eight more related Agta families who worked and shared in the harvest.

Woodburn (1982a:434) has classified foraging societies that have the following characteristics as immediate-return types: flexibility in social groupings; freedom of choice in social activities; lack of dependency on specific people or social structures beyond a small family unit; an emphasis on sharing and mutuality. This section has shown how these characteristics describe the Agta when they were still foragers.

Exchange Relations with Lowlanders

Agta enter into personal exchange relationships with lowlander families. These relationships sometimes are passed from generation to generation, and both sides know the history. In the Casiguran area, this system of trading partners is known as the *ahibay* system. These exchanges are similar to reciprocity exchanges, as they are governed by mutual needs rather than the market logic of supply and demand. The Agta provide lowlanders with

Figure 1.6. Agta hunter shares small portions of meat from two wild pigs with camp members. Larger portions are traded or sold to lowlanders, who have gathered around. (T. Headland, 1985.)

labor or forest products, especially meat, in return for some manufactured goods and agricultural products, especially rice, which is the Agta's staple food (see Figures 1.7 and 1.8). Griffin (1991:208) has further described the dynamics of the system.

> Special events in a partner's life may require major sacrifice from the other side. Weddings, funerals, and baptisms, illnesses, typhoons, crop failures, and periods of hunting failure all force the extension of credit and support. An Agta especially may need to be located at the farmer's home to work as a servant-helper in time of crisis. The farmer does not, as an economic and social superior, locate himself at an Agta homesite, except to be certain of obtaining resources as they come into camp, but often has to extend considerable credit. Agta often need carbohydrates when they have no meat to exchange. They may be ill, desire clothing, and especially crave tobacco. A good partner will give what he can to his Agta partners, thereby consolidating a sense of obligation, a debt, that favors his good treatment when the Agta have game or fish. In fact, the farmer's ability to take a generous portion in relationship to established rates of exchange is much greater if he has been generous in time of need. The old, long-lasting partnerships were often reasonably equitable, given the points of view of the foragers and the subsistence level farmers involved, and the intergenerational nature of the bonds.

History of the Agta

Early History

The contemporary Negritos are probably the descendants of aboriginal inhabitants of the Philippine archipelago going back at least 20,000 years. They appear to have been hunters and gatherers for much of their past history. Although their original language was non-Austronesian, their contemporary languages are Austronesian borrowed from Proto-Austronesian peoples who immigrated to the islands around 3000 B.C., the majority of whom were shifting swidden cultivators. Linguistic indicators imply that the ancestors of today's Philippine Negritos were in close contact with Austronesian-speaking farming communities well before the first century A.D. (Headland and Reid 1989, 1991; Reid 1987, 1994). This includes the ancestors of all of today's 10 Agta ethnolinguistic populations in eastern Luzon.

Figure 1.7. An Agta in a dugout canoe with outrigger on his way to exchange with his *ahibay*. (J. Headland, 1966.)

Figure 1.8. Agta woman pounds rice to remove husk. The rice was obtained from lowlanders. (T. Headland, 1980.)

The Spanish Period in Casiguran: 1578–1898

Magellan arrived in the Philippines in 1521, although Spanish coloniza-
tion did not begin until 1565 under Miguel López de Legaspi and continued for more than 300 years. Franciscan friars came to the Casiguran area
and began their evangelization of the Filipino population there in 1578.
They made contact with the Agta but had little success in converting them
because the friars could not sustain the contact and because of the passive
resistance by the Agta to moving to towns, which were centers of His-
panization.

  During the Spanish period the Agta maintained close client ties with their
patron lowlander farmers. Headland (1986:194–226) has found colonial
documentation clearly stating the Agta symbiotic relationship with Filipino agriculturists based on patron-client relationships. For example, a 1746
narrative states that the Agta "get resin, wax, . . . rattan, anahaw palm,
etc., for the Indians [Filipinos] and the Indians pay them either with food
or with glass beads or some kind of arrow or pots, etc." The document
later says that the Agta "make use of the Maguinoo [patron or master],
and by this we have entered [into relationships] with them. They become
relatives [with the townspeople] and they call themselves brothers with
those of the town and so with him who is their friend." The documents
from this period also describe some Agta planting their own fields to supplement their forest hunting and gathering activities.

The Agta in the American Period: 1898–1941

The American period in the Philippines began in 1898 with the United
States assuming the colonial role from Spain. Life in the Casiguran area
continued much as it had been previously. Americans made a few visits to
the area (Barry 1901, Worcester 1912). In 1911 Capt. Wilfrid Turnbull of
the United States Army arrived in Casiguran with orders to "bring the wild
tribe under government control." He established an Agta reservation on
163 hectares (403 acres) near the mouth of the Calabgan River, 23 kilometers (14 miles) southwest of Casiguran. Turnbull (1930:31) described his
strategy:

> In order to get control of the various groups—impractical so long as
> they were scattered—the Secretary of the Interior, the Honorable Dean
> C. Worcester, issued an order obliging the Dumagats [regional Fili-
> pino term for Agta] to take up residence on a certain tract of land
> which was later to be set aside by the Government as a reservation for
> the tribe.

In this way the Agta were to be turned into sedentary villagers and become farmers. The purpose of a school opened on the Calabgan reservation in 1912 was to introduce and integrate the Agta into the national culture.

Numerous government reports mention the problems of getting Agta to take up residence on the reservation and, for those who did, of getting them to stay there. There is reference to the use of chains and leg irons. Oral histories from the Agta concerning the 1930s recall deceiving Filipino soldiers: alternately showing up at the reservation for free meals while they played at farming for a time, and then later hiding from the soldiers who would hike to Agta camps to round up children for the school. The government was never able to permanently settle the Agta on the reservation (Headland 1985a).

At the reservation school Filipino and American history were taught as well as speaking and reading in Tagalog and some English conversation. In 1976 the Headlands found that 65 percent of those who had been of school age between 1921 and 1941 attended the school; most of these attended for a year or less and did not attain literacy in Tagalog. In the 1960s none could speak English except for a few words, and some had memorized songs. However, Headland believes that the school was the psychological seedbed for later changes. The Agta children had intense interaction with lowlander teachers and carried home new ideas to their families (Headland 1986:236–37).

In the late 1920s the Agta of Casiguran were visited by two U.S. Army pilots and a U.S. Navy survey ship. The U.S. Army was mapping the Philippines and then, as now, most areas of northeastern Luzon were accessible only by sea or air transportation. Lt. George Goddard was one of the survey aviators who spent several days in Casiguran, which served as a base for surveying the northeast coast of Luzon. The Agta assisted with the survey work. His account and photographs (Goddard 1930) show his fascination with the Agta.

World War II: The Japanese Period—1942–1946

World War II was a traumatic event for the Philippines. Much of Manila was destroyed. However, the semi-isolation of the Casiguran area spared it from much of the suffering and destruction that happened in other regions. The reservation school was closed upon the outbreak of hostilities. A platoon of Japanese infantry was stationed in the town of Casiguran. They treated both lowlanders and Agta well, and there was little interference with their customary ways. Some Agta interacted with the Japanese by developing exchange relationships such as they had with lowlanders, while others worked for the Japanese as guides, hunters, and casual laborers.

With the fall of Manila in March 1945, the Japanese forces retreated toward northern Luzon. In the mountains dividing the south-central Luzon and Cagayan valleys at Balete (Dalton) Pass, they made their last major stand. In June 1945, after five days of bitter fighting during which 10,000 Japanese were killed and 7,000 Americans killed or wounded, the Japanese were defeated. Thousands of Japanese soldiers retreated into the Cagayan Valley and from there eastward into the Sierra Madre Mountains. They were attempting to reach the northeast coast around Palanan, where they thought Japanese ships would rescue them. In their passage through the Ilongot tribal area in June 1945, Rosaldo (1980) estimates they directly or indirectly killed a third of the Ilongot population. From mid-1945 to mid-1946 starving and disorganized remnants of the Japanese forces were in the Sierra Madre Mountains and on the Casiguran coast roaming Agta forest areas.

The relative tranquility of the war period in Casiguran was disrupted on June 20, 1945, with the landing of 700 American troops who were joined by 300 Filipino guerrillas (Soeda 1985). Their mission was to mop up the Japanese survivors on the coast and in the adjoining mountains. None of the Japanese platoon stationed in Casiguran survived (Takamiya 1975:89–90). When the Japanese stragglers attempted to take food from the Agta, the Agta killed a number of them. Hundreds of other Japanese died from exhaustion and starvation. No Casiguran Agta were killed by the Japanese during the war. The Agta were well treated by the American forces, who dropped C rations by parachute when the Agta were isolated in their camps by the fighting. Later the Americans traded blankets to the Agta for baby monkeys. With the end of the war in September and the mopping up or surrender of the last Japanese soldiers in the area by mid-1946, life in Casiguran, including the peninsula, settled back to its traditional routine.

Summary

Until the 1960s the San Ildefonso Peninsula remained relatively secluded from national changes, and the Agta continued to live their traditional life by hunting and foraging in their forest habitat and exchanging some of the forest products for agricultural produce with the lowlanders. The reservation school at Calabgan, some minor logging in 1937–38, and the brief encounters with Japanese and American armed forces in the 1940s gave the Agta some experience of other groups. But there had been no stimulus or opportunity for structural change to take place, and most Agta continued their traditional ways. The next two chapters will show how this state of affairs was soon to change.

NOTE TO CHAPTER 1—USE OF GROUP TERMS

There is ambiguity in the meaning of the three ethnonyms used in this book: Agta, Filipino, and lowlander. This note attempts to clarify their usage.

Cultural Meanings

1. *Agta*—These are the 9,000 Negritos belonging to 10 distinct Agta linguistic groups who inhabit eastern Luzon. They have a distinctive physical appearance and different cultural forms than those of the Filipinos, although these differences are changing. This book is about the population dynamics of the San Ildefonso Agta, that is, the Agta residing on the San Ildefonso Peninsula. They are part of the larger Casiguran Agta linguistic group who speak a common language and who reside in the municipality of Casiguran. However, Headland has previously used the term "Casiguran Agta" somewhat differently. The boundaries of the municipality of Casiguran have been reduced at various times since 1966 because of the establishment of new municipalities (see Table 3.1). In earlier publications, Headland frequently used the term "Casiguran" to include the area of the municipality before these boundary changes. Headland's demographic survey of the Casiguran Agta (1986, 1989) refers to those Agta who speak the Casiguran Agta language as their first language regardless of where they resided. His usage excluded Agta in-migrants to the municipality who do not speak Casiguran Agta as their first language. In this research the term "Agta" will usually refer to the San Ildefonso Agta. But sometimes, especially in the first three chapters, it will refer to the Casiguran Agta living in the municipality and occasionally to all Agta. The context should clarify the usage.

2. *Filipino*—These are members of one of the eight major Christianized groups (Cebuano, Tagalog, Ilokano, Hiligaynon, Bikolano, Kapampangan, Pangasinan, Waray) and comprise 85 percent of the nation's population. The Agta are not Filipinos when the term is used in this cultural sense.

3. *Lowlander*—This is a subdivision of Filipino in the cultural sense. It usually refers to a rural Filipino, frequently working as an agriculturist or in some way dependent on agriculture. This term has its own ambiguity.

The lowlands were historically favored for agriculture, although agriculture is not restricted to the lowlands. Therefore, some people referred to as lowlanders may live in mountainous areas.

Civil Meaning

Based on the Philippine constitution, all minority tribal people in the Philippines, including Agta, are citizens of the Republic of the Philippines. All Agta are Filipinos in the civil sense but not in the cultural sense. In these pages we will usually use Agta and Filipino in the cultural sense. However, this usage does not intend to be misleading about the citizenship of the Agta. If there is departure from this usage, the context should clarify the meaning.

Part II

Loss of Regional Isolation

Chapter 2

Loggers, Homesteaders, Civil War, and the Market Economy

The previous chapter sketched the history of the Casiguran area, which includes the San Ildefonso Peninsula. The traditional Agta way of life was a part of this history. The Sierra Madre Mountains and the frequent storms of the Pacific Ocean provided physical barriers to transportation so that the Casiguran area was semi-isolated from the rest of the Philippines and not greatly affected by events taking place at the national or international levels.

This traditional way of life changed in the second half of the twentieth century. As the world became more connected because of the spread of the market economy aided by advances in transportation and communication, outside forces began to impinge on the Casiguran area. This chapter discusses the factors bringing about this loss of regional isolation. For each factor there will first be a discussion of the forces at the international and national levels giving impetus to social change, and then a description of how these changes came to the Casiguran area. These factors and their interaction have been incorporated into a model of deforestation developed by David Kummer (1991:96). Figure 2.1 reproduces Kummer's diagram of his model, which serves here as an outline and summary of the first part of this chapter. Deforestation was one result of these pressures impinging on the Casiguran area from the national level and in turn led to other changes, which are discussed in this and the following chapter.

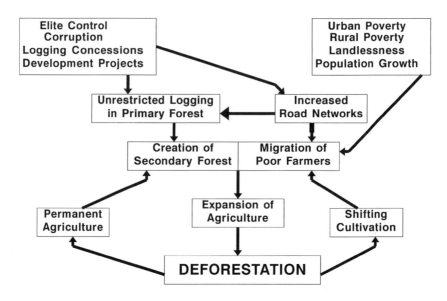

Figure 2.1. Kummer's model of Philippine deforestation.

THE PHILIPPINE TIMBER INDUSTRY

Its Importance in the Filipino Economy

The industrial world is completely dependent on nature's raw materials to feed its complex technology. For economic or geographical reasons, many of these raw materials are not produced in the industrial world. The Philippines is an archipelago of more than 7,000 islands, although only 800 are inhabited and only 11 have an area of 2,600 square kilometers (1,000 square miles) or more. These 11 larger islands comprise 90 percent of the total land area of the archipelago. They are covered by mountains and valleys and have a hot, moist tropical climate. It is an ideal setting for the production and extraction of raw materials to feed the appetites of industrial systems.

In 1989 lumber was the eighth most valuable export of the Philippines. According to government figures (Forest Management Bureau 1989:129), these exports were valued at approximately a quarter of a billion U.S. dollars. Japan, Taiwan, the United States, the United Kingdom, Australia, France, and Spain were some of the main destinations. These exports accounted for between 1 and 2 percent of the gross national product. Although the accuracy of all government figures with regard to Philippine forestry is open to question (Kummer 1991:104, 154, passim), it is clear that logging is an important source of economic activity in the Philippines.

Most of the logging in the Philippines is done in the uplands (slopes of 18 degrees or more), which comprise 55 percent of the Philippine land area. These are public lands, since land registration and ownership of land have historically been restricted to agricultural and urban areas, in other words, the lowlands. The timber industry is controlled by the government through a system of concessions to private companies that harvest the public lands. These permits state the geographical limits of the upland concession and the amount of timber that may be harvested based on the government's estimate of the sustainable yield of the area. The granting of these concessions has a continuing history of corruption (Kummer 1991:86, 70).

The Timber Industry in Casiguran Municipality

The search for and harvesting of raw materials eventually penetrated into northeastern Luzon and pierced Casiguran's semi-isolation. Prior to 1960 a few logging camps had operated in the municipality. In 1928 there was a logging camp at Dibakong, seven kilometers west of the town of Casiguran. It was owned by the Philippine Lumber Exportation Company with two Japanese managers and was probably of Japanese ownership. In 1937 it transferred its operations to the upper part of the San Ildefonso Peninsula and was a sizable operation for the next two years. In the early 1950s there were two other smaller camps on the lower half of the peninsula managed by a Spaniard and a Chinese, although both camps existed for only a few years.

In the 1960s there was an invasion of Aurora Province by the industry. By the late 1970s, 40 logging concessions were operating in the province, employing, by one estimate, 10,000 workers (Parumog 1982:2). Four of these concessions were in Casiguran: the Casiguran Bay Timber Co., the Industries Development Corporation, the RCC Timber Co., and the Pacific Timber Export Corporation (PATECO). These concessions opened large areas to the loggers since 85 percent of the municipality of Casiguran was government-owned timberland. Almost all the San Ildefonso Peninsula fell into this category (NAMRIA 1991:9 and Map 3).

The local headquarters of the concessions were located near Casiguran Bay, where earthen piers were constructed to bring in supplies and take out the logs by ship. Workers, frequently Agta, were employed to locate the harvest areas containing enough large-sized trees. Trucks and bulldozers set out from the local headquarters and used the beaches and riverbeds as

Figure 2.2. Bulldozers travel on the beach to site of log harvest. (J. Early, 1994.)

Figure 2.3. A truck bringing out a harvest of logs. (J. Early, 1994.)

Figure 2.4. The trucks deposit the logs on the beach prior to loading on ships. (J. Early, 1994.)

roads to approach the cutting areas (see Figure 2.2). Then bulldozers would cut a logging road from the riverbeds or from the beaches themselves to the harvesting area. Chain saws would topple the desired trees. Tree trunks would be cleaned of their branches and dragged by motor-driven cables to the trucks and loaded on them. The trucks would descend by the roads, rivers, and beaches to the headquarters site (Figure 2.3), where the logs were dumped on the beaches (Figure 2.4), pushed into water by the bull-dozers, and floated out into the bay. There they would be loaded on ships either directly for export or to be taken to Filipino sawmills. The logs were mostly from six species of the family Dipterocarpacae (Headland 1988a:125), which reach a height of approximately 40 meters (140 feet) with a buttress-trunk diameter at ground level of 6 to 8 meters (19.7 to 26.2 ft.). By the late 1970s, well over 90 percent of the mature trees of these six species had been removed from the Casiguran forests.

While there was no clear-cutting, the ecological impact of the logging process was severe. Taking the large trees removed the canopy of the rain forest. This action, combined with the noise generated by the trucks, bull-dozers, and chain saws, drove from the area many of the deer, wild pigs, monkeys, various bird species such as the hornbill (*Buceros* sp.), and other game. The widespread use of shotguns brought in by the loggers seriously depleted what game remained. The use of the streambeds as roads turned them into conveyor belts of mud that killed or drove off freshwater fish and shrimp. It created mud flows that spilled into the bay, smothering coral reefs and killing off much of the marine life. The loggers brought in dyna-mite, pesticides, and electrical prods, which they used to fish the coral reefs as well as the streams.

The Agta and the Loggers

The depletion of the closed-canopy forest, game, and aquatic life during the 1960s and 1970s destroyed the ecological basis of the Agta foraging way of life. However, the Agta found wage employment with the logging companies as guides, markers of trees to be cut, equipment guards, and unskilled laborers. They helped provision the loggers by hunting game (us-ing the loggers' guns) and further depleting their most important faunal resources. These new roles meant that the Agta were becoming wage la-borers in a market society with its distinctive logic of supply and demand, which was very different from the logic of reciprocity.

Many Agta welcomed the presence of the loggers, most of whom were friendly to the Agta in the traditional Filipino manner. From the Agta view-point, the loggers were wealthy and were seen as prospects to be good

Figure 2.5. A bulldozer driver distributes diesel to Agta women. (J. Early, 1994.)

ahibay partners. The loggers gave the Agta better exchange values for their meat than they received in town, free transportation on passing trucks, and gifts of diesel oil for their lamps (Figure 2.5). They provided the Agta with food and new goods in a quantity they had never had before. The Agta paid little attention to the long-range consequences of the ecological changes and what they might mean for the Agta's future subsistence.

MINING

In 1960 the Acoje Mining Company inaugurated an open-pit magnesium mine at Dinapigui, near the boundary between Aurora and Isabela Provinces, a two-day walk from the peninsula. The mining site was originally almost solid rain forest inhabited by an Agta band who were driven out by the mining company. The influence of the mine affected a wider area including the municipality of Casiguran and north to Palanan, Isabela. Large numbers of workers including many Igorots and Ilokanos in-migrated from other provinces to work at the mine. Headland estimates 200 to 250 Agta lived there at various times between 1961 and 1965. Agta laborers were paid the equivalent of one U.S. dollar a day. This was almost four times the standard day's wage in Casiguran at that time.

The mining operation was greatly reduced after 1965 and finally closed in 1972. But most of the lowlander mine workers remained in the region because there was no land available in the provinces from which they had come. Most became homestead farmers on land the Agta had previously used for their hunting and foraging.

PHILIPPINE POPULATION GROWTH AND INTERNAL MIGRATION

National population pressure was another factor that broke the semi-isolation of the Casiguran area. The post–World War II period in the Philippines was marked by a rapid population increase. Table 2.1 expresses this growth in three ways in columns 2, 3, and 7. Column 2 shows the absolute size of the Philippine population in various years. Column 3 expresses the growth as the percentage annual growth rate (r), and column 4 transforms this expression into the more intuitive form of how many years it would take the population to double if the growth rates of column 3 were sustained. Column 7 expresses the growth in the form of the crude rate of natural increase (CNI), which is the difference between the crude birth rate, column 5, and crude death rate, column 6. (When migration is negligible, columns 3 and 7 should be similar although they are based on different sets of data, slightly different time frames, and the difference between per 100 and 1,000 population.)

The size of the Philippine population grew from 7.5 million people in 1903 to 72 million in 1996. The pre–World War II period was characterized by a 2-percent rate of population increase, or a population doubling every 35 years. This rate of increase resulted from a high crude birth rate of approximately 50 and a high death rate of about 27, except for the epidemic years of 1902–4 and 1918–19 (see De Bevoise 1995 for the period 1876 to 1908, especially page 12). A slight decline in the nation's death rate took place in the 1930s, but this was interrupted by World War II in the first part of the 1940s.

The rate of increase climbed from 2 percent to over 3 percent with the sustained fall of the death rate beginning in 1945. This was due to advances in the prevention and control of infectious and parasitic diseases along with the growth in the delivery of health services after World War II. During the first part of this decline, from 1940 to 1975, the birth rate remained fairly constant. These birth and death rates are characteristic of the first phase of the classical model of the demographic transition. The change in the rate of increase from 2 percent to 3 percent meant that the population needed only 23 years to double instead of 35. (This is popu-

Table 2.1. Size and Rates of Increase of Philippine Population, Crude Birth and Death Rates, 1885–1995

Year	Population (thousands)	% Rate Inc. (r)	Years to Double	CBR	CDR	CNI
1	2	3	4	5	6	7
1885–89	—	—	46.2	—	34.8	15.2
1890–94	—	—	57.8	—	37.8	12.2
1895–98	—	—	38.0	—	32.4	17.6
(1902–4)	—	—	—	—	58.0*	-8.0
1903	7,635	—	—	—	—	—
1918	10,314	2.0	34.7	—	—	—
(1918–19)	—	—	—	!=50	!(47.0#)	—
1940	16,459	2.1	33.0	45.3	!=27.0	!=23.0
1945	18,083	1.9	36.5	—	—	—
1950	20,275	2.3	30.1	—	—	—
1955	23,568	3.1	22.4	45.3	19.2	26.1
1960	27,377	3.0	23.1	44.8	16.5	28.3
1965	31,770	3.0	23.1	44.2	14.2	30.0
1970	36,852	3.0	23.1	44.2	12.2	32.0
1975	42,071	2.7	25.7	41.0	10.3	30.7
1980	48,098	2.7	25.7	36.2	8.6	27.6
1985	54,378	2.5	27.7	33.3	8.4	24.9
1990	60,684	2.2	31.5	34.1	12.5	21.6
1995	68,400	2.1	33.0	30.0	9.0	21.0

* cholera epidemic
# influenza epidemic

The vertical bar with an equals sign beside it means that the crude rate after the equals sign covers the entire period marked by the vertical bar.

Column 3: percentage annual growth rate
Column 5: crude birth rate
Column 6: crude death rate
Column 7: crude natural increase (column 5 minus column 6)

Sources: This table has been compiled from United Nations sources (Demographic Year Books [DYB], United Nations 1978 [UN], and the Population Reference Bureau) except for the population in 1903 and 1918, which are census figures. Philippine demographic data are deficient in many respects (United Nations 1978:338–49). Filipino demographers and the United Nations Population Division have attempted corrections of enumerated data, and these have been used wherever possible in the table. The table is important for trends rather than exact values.

Populations
Censuses 1903–75, UN:11
Estimates 1932–47, DYB 1949–50:100
Estimates 1948–78, DYB, Historical Supplement to 1979:145–46
Estimate 1980–85, DYB 1985:137, 147
Estimate 1985–90, DYB 1990:120
For 1995, Population Reference Bureau 1995

Crude Birth and Death Rates
1879–98, UN:346
1903–60, UN:117
1950–74, DYB, Historical Supplement to 1979:145–46
1950–70, UN:348
1946–73, UN:347
1975–80, DYB 1981:325
1980–85, DYB 1985:147
1985–90, DYB 1990:120; DYB 1992:120

Crude Death Rates
1920–73, UN:100–102

larly known as "the population explosion.") In the early 1970s fertility also began to decline from around 45 to 34 per 1,000 population in 1990, slowing the rate of increase to about 2.4 percent or a rise to 29 years for the population to double. This marks the beginning of the second phase of the classical transition model. This postwar growth of 42 million people in a little under 50 years has posed the pressing problem of how to provide increased resources for the increased population.

National Development to Provide for Population Growth

The various strategies of economic development are usually proposed as the answer to this problem. If the success of these strategies can be measured by the average rise in per capita gross national product, then the Philippines appears fairly successful. However, the per capita average is misleading. Figure 2.6 shows the increase in the average per capita gross national product as well as the decline in real wages that accompanied this growth. The top 10 percent of Filipino families controlled 37 percent of the nation's income in 1985 while the top 20 percent controlled almost 60 percent of the nation's income at about the same time (Kessler 1989:18). Kummer (1991:80), citing Mangahas and Barros, notes that in 1975, 60 percent of the population was living in poverty. As Kummer (1991:91) concludes, "The major process that has occurred in the past 45 years in the Philippines has been the impoverishment of its people. Much of the expansion of agriculture and migration has been the result of desperately poor people trying to find a place where they can at least feed their families."

Migration

The population-resource pressure in the Philippines has created streams of migrants seeking a solution to their problems. Other than the expansion of metropolitan Manila, the starting point of most migration has been the rural areas. Land is unavailable, which prevents the expansion of peasant small holdings. There have been two main types of migration. One stream seeking unskilled work in the expanding commercial sector has headed to urban areas, resulting in a surplus labor supply and extremely low wages. The other stream consists of homesteading agriculturists who have gone from province to province seeking any available land to raise subsistence and market crops. Most available sites are public uplands owned by the government. These are mountainous areas that originally had no roads to get into the potential agricultural sites or to take commercial crops to market. The usual solution has been to migrate to an upland area after loggers

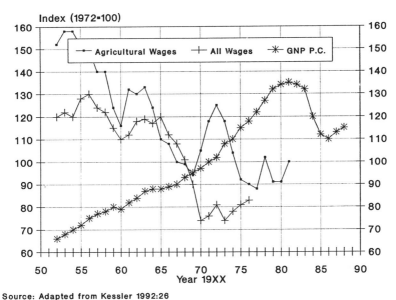

Source: Adapted from Kessler 1992:26

Figure 2.6. Trends of real wages and national income.

have finished harvesting timber. When the loggers and their machines de-part, they leave behind the roads they have cut. Their removal of trees also means that the work of clearing the land has already begun.

DEFORESTATION

This interaction of logging, migration, and the expansion of agriculture in the uplands of the Philippines has created a serious problem of deforesta-tion. Around the turn of the century observers estimated that 70 percent of the Filipino land area was covered by forests. By 1950, when systematic studies began, this figure had declined to 50 percent. By 1975 it was down to 30 percent, and in 1987 it had reached 24 percent (Kummer 1991:45, 46, 56).

Deforestation has reached the rain forest of Casiguran. Before World War II, the area was 80 percent covered by old-growth forest. Because of the intensity of logging activities in the 1960s and 1970s, the municipality of Casiguran prohibited all logging in 1978, and the ban remained in effect until the mid-1980s. Despite the ban, logging continued because of bribery of some local officials. By 1990 Casiguran was 60 percent logged-over re-sidual open forest, 21 percent farmland, 10 percent deforested brushland or grassland, and only 9 percent remained as full closure forest (NAMRIA 1991:52). During our fieldwork in 1994 we observed intense logging in the Kinabunglan Valley on the western side of the peninsula midway between Pasaruboy and Dinipan (see Maps 1 and 4). This logging was illegal, according to investigators on the scene from a conservation group in Manila called the Philippine Association for Intercultural Development. Presumably they were correct, because with their arrival nine bulldozers and at least ten logging trucks were hidden in the forest.

LOWLANDER IN-MIGRATION AND OCCUPATION OF LAND USED BY AGTA

In-migration

The Casiguran area had the prerequisites for a large-scale in-migration of Filipino peasants. The loggers had been there since 1928 and in full force since the 1960s, which meant that there were a number of logging roads opening up the mountainous terrain. Lowlanders in search of land began in-migrating into the Casiguran area in the 1960s when there was an almost 10 percent annual rate of population increase. The influx was interrupted from 1974 to 1976 because of guerrilla warfare in the area, discussed later in this chapter. In-migration resumed in the late 1970s and has continued into the 1990s, although at a reduced rate because there was less land available.

Occupation of Land Used by the Agta

This migration has resulted in lowlanders occupying areas that the Agta had historically used for their campsites and foraging activities. On the San Ildefonso Peninsula the occupation followed a pattern that was typical for the region. The first lowlanders in an area would occupy public lands near the bay and as close to the town of Casiguran as possible. Small boats solved the transportation problem, so there was little dependence on logging roads. This land tended to be less hilly and easier to work. As these sites became occupied, the new arrivals began to use the logging roads to gain access to the interior of the peninsula and establish their agricultural plots there. This was the area where the Agta traditionally had a number of their campsites. Headland (1986:238) notes that in the 1960s and early 1970s Agta camps were in a range of 20 to 120 minutes' walk upriver from lowlander sites. But by 1982 the Agta had lowlanders living next to or upriver from them. Logging and homesteading had penetrated into the heart of the rain forest itself, the traditional home of the Agta.

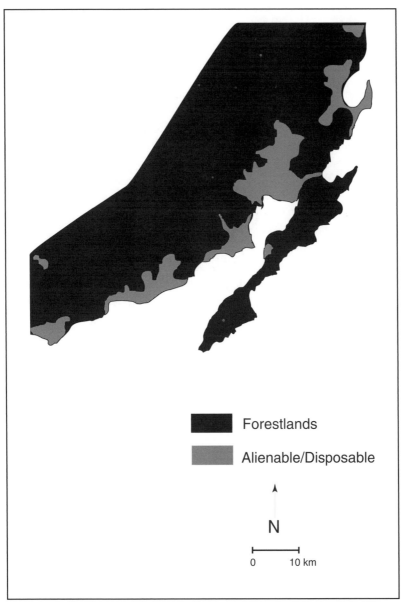

Map 3. Legal Status of Land in Northern Aurora (based on National Mapping and Resource Information Authority, 1991:58).

Lowlander and Agta Conception of Land Rights

Lowlanders perceived their occupation and use of this land as perfectly legal. A 1989 survey map (Map 3) estimated 85 percent of the land in the municipality of Casiguran and almost all the San Ildefonso Peninsula were public forest land under government jurisdiction (NAMRIA 1991:8, 9, 57, 58). The government allowed occupation and use of this public land. After a certain period of use, the occupant could apply for and receive a title giving ownership of the land. A land title was the goal sought by the migrating Filipino peasants to alleviate their plight. Since the Agta were not perceived as permanently present on much of the land nor as farmers, the Agta foraging areas were viewed as empty public lands.

The Agta did not view the lowlander incursion into the rain forest as an invasion of their land. From the Agta point of view it would be incorrect to speak of "Agta land." For the traditional Agta, there is no ownership of land in the legal sense by anyone whether it be lowlander, Agta, or government. The Agta see land as an unlimited resource since historically it has always been available. As long as an Agta is not actively using the land at the time, they usually register no protest when a lowlander begins to use it for agriculture. The Agta see the lowlander as establishing an understandable claim to the results of his labor the same as any Agta would.

However, they do not perceive this claim as giving ownership of the land itself. Traditional Agta society is not a market society, where any object including nature itself (land) and human beings (labor) can be owned and turned into a market commodity with a monetary value put on it by the blind forces of supply and demand. (This latter logic is characteristic of laissez-faire markets, which predominate in frontier areas.) In the traditional Agta view, anyone can use available land for whatever purpose desired—hunting, temporary swidden, planting coconut trees, and so forth. After the Agta have hunted or foraged in a certain area, they may go to another area. The concept of settling down on a piece of land, seeking ownership by legal title, and permanently working it is different from their cultural logic. (This does not mean to imply that the Agta have no rights to their traditional home areas. For an international legal perspective, see Bodley 1990:77–78 and the references cited there. For the Philippine legal situation, see Lynch 1983, especially page 462.)

These views of the situation do not imply that there were no conflicts between the Agta and lowlanders over land use. Some lowlanders disdained the Agta and simply seized land that they wanted regardless of its being used by an Agta. Headland (1986:426–38, 584–92) has collected a number of cases of these disputes, some of which led to homicides against the Agta.

A Road Reaches Casiguran

In 1977 the government constructed a single-track dirt road from Casiguran to the provincial capital, Baler, in southern Aurora Province. It could be negotiated by four-wheel-drive vehicles (usually weapons carriers) and motorcycles. At Baler it connected Casiguran with the national road system and Manila. The road allowed greater access to the national market and provided an easier entrance for an increasing number of in-migrants.

Civil War

The Nation

The Philippines has a long history of dissent and rebellion—over 300 years against the Spanish and later against the Americans (Constantino 1975). Independence at the end of World War II did not change social structures. The Americans facilitated the continuing concentration of resources and power into the hands of a wealthy elite who inherited control of the political structure at the time of independence. Kessler's (1989:18) analysis cited previously has indicated the extent of this concentration. It specifies some of the variables in Kummer's model (Figure 2.1), illustrating the stranglehold of the oligarchy in the deforestation crisis.

In an agricultural society the major form of capital and wealth is ownership of land. The concentration of resources resulted in increased large land holdings that in many cases became quasi-independent political entities with private armies. Peasant families became tenant farmers with patron-client relationships between themselves and the landowners. The landowners attempted to convert to a labor market governed by supply and demand in the 1930s and thus to forgo their quasi-reciprocity obligations to tenants. This change led to peasant unrest and resistance. In central Luzon these peasant groups were an important part of the guerrilla campaigns against the Japanese during World War II, and this activity evolved into the Huk protest and armed rebellion against the national government from 1945 to 1956 (Kerkvliet 1977).

These structural problems of Philippine society were heightened after the election of Ferdinand Marcos to the presidency in 1968. To remove all political opposition and create his dictatorship, Marcos declared martial law in 1972. Filipinos were enraged. A number of students, peasants, and intellectuals sought a solution in the Communist Party of the Philippines (CPP) and its military wing, the New People's Army (NPA). After its own internal struggles, the party had reorganized in 1969 and continued its resistance to the established system. From its headquarters in Tarlac, the NPA formed cadres and sent them to other parts of the archipelago to recruit for the revolution and initiate armed rebellion. The NPA needed an area to train these recruits, another Yenan following Mao's model. The remoteness of northeastern Luzon provided ideal conditions. In 1970 the NPA established a camp in the western foothills of the Sierra Madre range in southern Isabela (Jones 1989:48–52). This was only 50 kilometers (30 miles) north of Casiguran.

Martial Law in Casiguran

Under the provisions of the 1972 martial law, all firearms were confiscated. As a result, no guns were available for hunting what little game was left. The bows and arrows of the Agta were also confiscated. The Agta were later given permission to resume making bows and arrows, but little hunting was done with them owing to their ineffectiveness in conditions of game scarcity.

The NPA in the Municipality of Casiguran

As part of their expansion strategy, the NPA attempted to establish themselves in northern Aurora and Isabela. The NPA's Third Red Company entered northern Aurora in 1973, and by 1974 they controlled much of the region except the four towns of the four northern municipalities (Map 2). In the municipality of Casiguran, the government confined all lowlander civilians to the town itself under a dusk to dawn curfew. At night they remained inside their houses without any lights. Many Agta voluntarily came or were collected by the authorities into camps close to the town of Casiguran, where they were under the protection of the Philippine military. Toward the end of 1975 the military moved the Agta to the defunct reservation at Calabgan. This was another effort by the government to settle the Agta there and get them to take up agriculture (Headland 1985a).

In early 1975 the government moved the elite 55th Battalion of the Philippine Constabulary (PC) into the area and appointed a PC major as mayor of Casiguran. Armed clashes with the NPA ensued near each of the four towns as well as aerial bombardment of suspected NPA positions in the Sierra Madre, just upriver from Casiguran town. According to Jones (1989:55), the Third Red Company of the NPA, trapped in Aurora Province, split into two groups and began retreating in March 1975. One group of 55 officers and fighters fled westward, walking more than four months and fighting nearly 50 engagements with pursuing constabulary rangers before stopping in Nueva Ecija province 80 miles away. An entire detachment of guerrillas left behind in Aurora to defend a few hundred communist civilian supporters who chose to stay in the province was killed, and about 100 civilian supporters died.

Several dozen guerrillas and 200 civilian supporters escaped to southern Isabela. The PC remained in the Casiguran area through 1976.

The Casiguran Agta became involved in the conflict, working for both sides. Twelve men, nine women, and one child lived with and assisted the NPA for various periods of time as guides, hunters, and laborers. Two of the Agta men were wounded when they assisted in attacks on PC camps. Another was executed by the PC when he was caught trying to steal rice from a warehouse. A six-year-old Agta child from Casiguran was killed in 1979 in western Isabela in a shoot-out between the PC and the NPA group with whom her family was traveling. Another Casiguran Agta died while a prisoner of the military. He had been captured in the Palanan area while hunting game for the NPA. It is doubtful these Agta understood the extent of the conflict or the ideological issues that started it. Those whom the Headlands later interviewed said they traveled with the NPA for food and monetary rewards, for adventure, or for fear of NPA threats if they did not help them (Headland and Headland 1997).

Agta also worked for the PC as guides, hunters, and laborers. None were killed by the NPA. The government organized a paramilitary unit, the Civilian Home Defense Forces (CHDF). Many young Agta men joined. They were armed with M-1 rifles and assigned to guard duties and jungle patrols. Several were involved in skirmishes with the NPA, and at least two were rewarded for killing NPA guerrillas. As soon as the PC left the area, the Agta left the reservation. These experiences with the PC and NPA involved the Agta in peasant guerrilla warfare. It was an intensive learning experience about Filipino society into which they were being more closely incorporated at the bottom of the peasant hierarchy.

Economic Depression

The Aquino Assassination

In the early 1980s the nation's GNP (gross national product) had begun to decline. Because of governmental corruption, both the national and international business communities lost confidence in the Marcos government and economic activity declined. In August 1983, a popular opposition candidate, Benigno Aquino, Jr., decided to return to the Philippines after a three-year exile. When his plane landed, a military escort boarded it and shot him. This event created a deep sense of outrage among the Filipino people. The assassination became the focal point of a national crisis and an economic depression in the already fragile Filipino economy. The peso declined in value and inflation took hold. It eventually led to the overthrow and exile of Marcos himself in 1986.

Economic Depression in Casiguran

Since Casiguran by this time had lost its semi-isolation and was now part of the national economy, these events were quickly felt by the Agta. In Casiguran the price of rice rose 108 percent in the 11 months following the assassination while wages received by the Agta rose only 50 percent—a decline of 27 percent in real buying power. Rattan, which had become the major source of income for the Agta from 1980 to 1984, had been so overharvested that the Agta were unable to collect as much as previously, and there was a drop in collectors' incomes of 68 percent (Headland 1985b).

SUMMARY

This chapter has examined the history of the gradual loss of physical and cultural remoteness for the San Ildefonso Peninsula and the wider Casiguran area. This semi-isolation had been produced by its geography—the Sierra Madre mountain range quickly falling away to the Pacific shores, which were frequently whipped by dangerous northeast monsoon winds. For the first half of the century, the area was sparsely populated, with the few lowlanders living mostly in the towns that served as governmental centers for the municipalities. The mountainous forest areas were roamed by small groups of Agta for whom foraging was a way of life.

Around 1960 the centuries-old economic and cultural remoteness was broken by the greatly expanded efforts of the mining companies and logging companies seeking raw materials for the export market. This intrusion disrupted the ecology of the San Ildefonso Peninsula. Loggers destroyed the tree canopy of the rain forest and drove off the game animals; the few remaining game were depleted by the introduction of shotguns. Because the streams were used as roads, they became mud runs that destroyed their own aquatic life. The introduction of dynamite, poisons, and electrical stunners damaged the fishing grounds and marine life of the bay. This ecological upheaval destroyed the basis of traditional Agta subsistence.

Using the roads made by the loggers, the next group of in-migrants were homesteading Filipino peasants trying to cope with an expanding population. They took over land that the Agta had historically used for their hunting and gathering and in some cases attempted to take land under cultivation by the Agta.

Guerrilla warfare came to the area in 1974, and numerous Agta were actively involved on both sides. Many of them were relocated to the town of Casiguran and then to the reservation at Calabgan. Most Agta became dependent on wage labor for much of their subsistence. In 1977 a dirt road was constructed linking Casiguran with the provincial capital of Baler and with the national road system. In 1983 the Agta fell victim to the economic inflation and depression set off by Marcos's depredations. These experiences were periods of intense learning by the Agta about the Filipino world into which they were being incorporated.

Chapter 3

Cultural Change
for the Agta

Chapter 1 described the traditional way of life of the Agta Negritos in northeastern Luzon. Chapter 2 recounted how the national and international worlds have impinged on the region where the Agta of the San Ildefonso Peninsula live. This chapter will examine how these Agta have reacted to the loss of their regional remoteness and the resulting changes that have occurred in their lives in the latter half of the twentieth century.

THE GROWTH OF THE LOWLANDER POPULATION ON THE SAN ILDEFONSO PENINSULA

The demographic composition of both the peninsula and the rest of the Casiguran region changed dramatically with the loss of semi-isolation beginning in the 1960s. Table 3.1 shows that in 1960 there were about 5,000 lowlanders in the municipality of Casiguran (Map 2) with 86 percent of them residing in or close to the town itself. By 1994 the municipality population had grown to over 20,000. By 1990 only 41 percent of the lowlander population lived in or close to the town of Casiguran. The rest lived in the northern, southern, and eastern parts of the municipality, including the peninsula. The western part of northern Aurora is relatively unoccupied because it consists of uplands unfit for agriculture owing to slopes of over 50 degrees.

Until the 1920s the Agta were the only people living on the San Ildefonso Peninsula. Table 3.2 shows the change. In 1942 there were 215 lowlanders

Table 3.1. Lowlander Population of the Municipality of Casiguran, 1948–94, with Boundaries as of 1970

Year	Population	% Annual Increase
1948	3,047*	—
1960	5,066*	4.3
1970	12,128	9.1
1975	11,670	-0.8
1980	13,925	3.6
1985	16,616	3.6
1990	18,388	2.1
1994	22,000	4.6

*The boundaries of the Casiguran municipality were changed in 1966 by dividing off from Casiguran the new municipalities of Dilasag, Dinalongan, and Dinapigui. Therefore the populations in 1948 and 1960 for the area of the current municipality of Casiguran are estimated. For 1970 and 1980 the sum of all four municipalities was calculated and 54% of this sum used for Casiguran. The estimates were obtained by applying this percentage to the pre-1970 totals. (The total population in 1994 for all four municipalities in northern Aurora—the original Casiguran municipality until its subdivision in 1966—was 50,000.)

Sources: Censuses of the Philippines, 1980 and 1990, Dept. of Aurora, p.1. 1994 estimate by municipality officials.

and 179 Agta, an almost 1 to 1 ratio. There was very little increase in the lowlander population on the peninsula until the late 1970s, when the government declared the land open for homesteading. By 1994 there were more than 2,500 lowlanders and 231 Agta on the peninsula, for a ratio of more than 12 lowlanders for each Agta. This dramatic change of demographic density is a consequence of the changes discussed in the last chapter. It has brought alterations in the Agta way of life.

NEW FORMS OF SUBSISTENCE

Wage Labor for Loggers

Logging continued on the peninsula in 1995. The loggers in their search for large trees hired Agta to serve as guides in the rain forest. They also hired Agta as unskilled laborers in the harvesting process. During the periods when loggers parked their machinery in the forests, Agta were hired to guard the equipment.

Table 3.2. Lowlander and Agta Populations of the San Ildefonso Peninsula, 1942–94

Year	Population			
	Total	Lowlanders[1]	Agta	Ratio (L:A)
1942	394	215	179	1.1:1
1975	429	270	159	1.7:1
1980	1,105	916	189	4.8:1
1990	2,271	2,048	223	9.2:1
1992	2,510	2,305	205	11.2:1
1994	2,731	2,500	231	11.8:1

Source: Agta Population—this study

Lowlander Population
1942—estimate by Headland from number of households with estimated 5 persons per household based on census figures
1975—Census of the Philippines, Dept. of Quezon
1980—Census of the Philippines, Dept. of Aurora
1990—Census of the Philippines, Dept. of Aurora, p. 1
1992 and 1994—estimates by local officials of lowlander population including people in the Acculturating population

Note 1: Government census enumerations do not distinguish lowlanders and Agta. But government census takers in many places ignore Agta. The enumerators don't hike into the forest to find them or they consider the Agta as transients (cf. Rai 1990:127). Therefore census figures are interpreted here as lowlanders, although probably a few Agta were included. For the purposes here, the few Agta can be considered as compensating for some probable underenumeration of lowlanders.

Wage Labor as Agricultural Workers

The lowlanders pay Agta to do whatever work is demanded by the agricultural cycle for any type of crop, especially rice and coconuts. Agta have traditionally done this type of work but within the context of the ahibay system. Working as a wage laborer or sharecropper is much more impersonal, and there is little obligation beyond the work agreement itself.

Sellers of Forest Products

For centuries the Agta have collected forest products to exchange with farmers. With the penetration of the market system into this area, the Agta collect such products for sale to traveling traders, local lowlanders, or to people in town or at the logging companies. Among the principal forest items the Agta collect for commercial exchange are:

1. Rattan—During the 1970s there was a boom in the international market for rattan furniture. The Philippine export market sought rattan all over the islands. Beginning in 1979 rattan buyers arrived in Casiguran and often lived in camps with Agta in order to buy as much rattan as the Agta could harvest. Often the buyers would take Agta by motorboat to inaccessible areas where rattan was plentiful. By 1986 the rattan was almost depleted from overharvesting.

2. Orchids—In the mid 1980s some Agta began harvesting wild orchids for sale to traders in Casiguran, who take them to Manila for resale.

3. Imperata grass and leaves of the livistona and nipa palms—These are sold to lowlanders as thatch for roofing houses.

4. Wooden poles—These are sold to lowlanders for house posts and other building parts.

5. Firewood—Sold to lowlanders for cooking purposes.

6. Wild Honey—Sold to lowlanders.

7. Wild game and fish—Over half is sold to lowlanders.

Joint Venture

Another way the Agta have learned to cope with their changed conditions of subsistence is having a joint arrangement with a lowlander. The lowlander gives the Agta a young domestic pig or a fishnet, or loans him a gun. The Agta raises the pig or fishes with the net or hunts with the gun. When the results of the effort are collected or sold, the proceeds are divided, usually equally, between the Agta and the lowlander.

Agta as Sedentary Agriculturists

Some Agta understand the legalities of homesteading and have acquired legal ownership of land. One family on the peninsula and two on the mainland have become full-time rice farmers in a manner similar to their lowland peasant neighbors.

Agta Girls Working as Maids

Throughout the Philippines, many middle-class households have maids to do menial work. The maid lives in the household while the family provides her food, clothing, and any necessities in lieu of paying her a salary. Before 1950, two Agta girls from the peninsula had worked as maids. Between 1950 and 1994, 33 Agta teenage girls had left the peninsula to live in low-lander households for this purpose. Frequently they went to Manila. Most of these girls eventually return to their families. But as will be seen later, their learning experiences from living in Filipino homes in distant cities has had an acculturating effect.

TIME ALLOCATION OF THE AGTA

The changes described above in Agta subsistence patterns are dramatically shown in the time that two Agta populations spent in hunting in the early 1980s. Table 3.3 is constructed by reorganizing tables by Rai and the Head-lands to make them comparable. Rai (1990:179–80) studied a Disabungan Agta group in the province of Isabela, which at the time of his study (1980) still lived the traditional Agta lifestyle based primarily on hunting, fishing, gathering, and trade of forest products for rice. Headland (1986:310–33; 468) studied 331 Agta adults over a 19-month period in 1983–84 in Casi-guran (including the peninsula) after they had been forced to discontinue traditional hunting and gathering and were becoming peasantized. Table 3.3 shows the dramatic decrease in time allocated to hunting as Agta groups change from foragers to ex-foragers. The Disabungan Agta studied by Rai live in an interior mountainous area 64 kilometers north of Casiguran. They were thus able to retain their traditional foraging way of life a few years longer than the Casiguran Agta, who have more contact with their proximity to the sea. Thus the two groups are not fully comparable in their remaining activities.

NEW RESIDENCE PATTERN

The changes of subsistence patterns and demographic density have resulted in a different type of settlement pattern. Most Agta of the peninsula no longer live in camps solely composed of kin-related nuclear families. Instead the increasingly common pattern is a nuclear family residing in proximity to a lowlander farm or lowlander settlement that provides work for the Agta. The exchange may be conducted on a weakened patron-client

Table 3.3. Time Allocation for Various Activities for Two Agta Groups in the Early 1980s under Forager and Peasant Conditions*

Activity	Male Agta		Female Agta	
	Disabungan Forager	Casiguran Peasant	Disabungan Forager	Casiguran Peasant
Hunting	75%	8%	9%	1%
Fishing	3%	8%	6%	8%
Swidden	—	8%	1%	12%
Travel	15%	8%	33%	18%
Equipment making	—	5%	—	8%
Gathering	1%	43%	25%	36%
Wage labor	—	20%	—	18%
Other	7%	—	27%	1%
Total	100%	100%	100%	100%

* Data on Disabungan Agta in Isabela Province for 1980, calculated from Rai 1990:180. Data on Casiguran Agta for 1983–84 calculated from Headland 1986:468. Both data sets are recalculated to make them comparable.

basis, but impersonal wage labor has been the more common form since the 1980s. The Agta house is usually a few minutes' walk from a lowlander residence. The house style has also changed. In the 1990s many Agta obtained milled boards left by loggers and used them for the sides and floors of their small dwellings. This is copied from the style of the peasant houses of the lowlanders.

THE NEW TYPE OF SOCIETY

The changes in the Agta way of life that have taken place since the 1960s have resulted in a change of societal type for the Agta. The traditional Agta man was a hunter of game in the rain forest with bow and arrow. Although hunting was the point of pride for the Agta and formed an important part of male identity, gathering was also important, especially for the women. Meat and forest products were exchanged with lowlander agriculturists mainly for rice, the Agta staple.

In the new situation, the Agta do little hunting. Most gathering is for items that have commercial value in a market economy. Their patron-client arrangements with lowlanders have been greatly weakened, and wage labor has taken their place. Owing to the ecological transformation of the area by loggers and agriculturists, the Agta can no longer obtain the types and quantities of forest products they used to exchange with trading partners. In order to obtain rice, they must work for a small share of the harvest or buy it with cash gained either by the sale of their labor or of any item they can collect or produce.

The changes have not been restricted to how they make a living. The lowlanders have brought in not only the national economy but also the legal system, medical services, and schooling. Very few of these public services reach the Agta, and if they do it is usually on a last and least basis. The Agta are being incorporated into rural Filipino culture. Although they retain a distinct ethnic and linguistic identity, there is a systemic change of their previous foraging style of life. An examination of rural Filipino stratification helps us to understand this change.

RURAL FILIPINO STRATIFICATION

Filipino peasants are not homogeneous. Anderson's (1964) study of rural Filipino stratification based on access to land includes in its various categories several strata of peasantry and is the basis of the discussion here. His strata of peasantry agree with the neo-Marxist threefold class division of peasantry (Wolf 1969:289–94) but expands these classifications and gives details of the Filipino scene.

Owners of Large- and Medium-Sized Land Holdings

These landlords usually live in the provincial towns or in Manila. Their ownership of land gives them power in the rural sector. Peasants provide the agricultural labor for their lands. Large land holdings are a common pattern in some regions of the Philippines, although Casiguran has few. Some homesteaders have acquired medium-sized holdings and do no agricultural work themselves.

Owners of Family-Size Holdings

Family-size holdings are of moderate size so that the holding could sustain a family by its own efforts without requiring outside labor. The only exception may be some additional help at peak periods in the agricultural cycle. The owners of these holdings are relatively secure since the size of their holding guarantees subsistence. They may engage in market activity when conditions are favorable, but few would give up the security provided by subsistence farming for the risks of the agricultural market. There are three subtypes of this category: farmers-owners, nonfarmers-owners, and farmers-tenants. Farmers-owners work their holdings themselves. Nonfarmers-owners do not work their holdings themselves but have someone else do it, usually tenant farmers. Farmers-tenants work as tenants on someone else's land in addition to working their own small holdings.

Tenants

These families may have small holdings of their own, but they primarily work the land of others. This work is not wage labor but is performed under an implied set of quasi-reciprocity rules (Lynch 1970; Hollnsteiner 1970). These rules are not set down in writing but are simply the culturally accepted ways of doing things in the rural Philippines. The tenant exchanges his labor on the land for a portion of the harvest and the right to request and to receive loans and various kinds of assistance from the landlord when the tenant is in need. This is the tenant's security, and the community negatively sanctions failure to honor these obligations by a landlord. In central Luzon during the 1950s, landlords attempted to substitute the market logic of supply and demand in place of tenancy by restricting the tenants to wage labor without any further social and economic obligations entangling the landlord. This rejection was one of the main factors behind the peasant Huk rebellion (Kerkvliet 1977).

Landless Agricultural Workers

Families of this type usually exchange their labor in various types of occupations for wages or food, anything in an attempt to provide for their subsistence. Many are underemployed and suffer frequent unemployment. They are at the bottom of the rural hierarchy because they lack the two basic securities of rural life—ownership of land and/or a tenant arrangement whereby a land owner takes some responsibility for their well being.

Salaries from Nonagricultural Work

This group falls outside the above classification because their work does not involve access to land. It is difficult to equate this group in social prestige with any of the above because it depends on the level of income received.

THE AGTA AS FILIPINO PEASANTS

With the basis of their foraging way of life destroyed in the 1980s, the San Ildefonso Agta have been incorporated into rural Philippine society in one of the three peasant strata outlined by Anderson (see also Griffin 1994). There are a few exceptional cases where Agta have become owners with land titles of family-size holdings and work the land themselves growing rice.

Others would be classified as tenants because of their extensive and continuing ahibay relationships with lowlanders. They mostly trade their labor to their trading partners. They may have access to land they have traditionally used for small swidden plots or to grow some fruit or coconut trees. As long as Agta continue to use this land, their occupancy is recognized by both Agta and those lowlanders not attempting to exploit them. For the lowlander, this recognition is not dependent on any written documents but on customary or living law. These "holdings" are usually small since they do not include the larger traditional foraging areas that are seen by most lowlanders as unoccupied and subject to homesteading. Some Agta retain the mind-set of their foraging days and prefer to rely on their ahibay relationships rather than attempt to become full-time agriculturists.

Most Agta under the changed conditions belong to the lowest stratum, landless agricultural workers. They have lost the two foundations of their traditional subsistence: access to land and ahibay relationships with lowlander families. They have lost access to land, either because it has been taken by the homesteading process when "unoccupied" or by coercion. They have lost ahibay agreements because lowlander families who have lived in the region for some time have dropped the custom and many of the in-migrating homesteaders do not understand the custom with reference to the Agta. Beginning in the 1960s, and especially since 1980, most Agta have changed from being foragers to becoming part of the rural labor market working for the lowlanders—those long resident in the area as well as homesteaders. In the Casiguran region a number of homesteaders have become owners of medium-sized holdings. Many live in the town and do not work the land themselves but rather hire Agta to do so. On the peninsula the homesteaders tend to have family-size holdings but still are in need of intermittent Agta labor.

This stratification system can be seen not only in terms of status but also class conflict. Anderson (1964:175) notes:

> At the top of the hierarchy . . . status difference becomes consolidated into social class difference, with more stringent mechanisms inhibiting mobility and maximizing cultural differences and social distance. At this level persons in each class have complementary functions the more important of which, over the long term, being contributed by the lower segment. It is advantageous for the upper class to perpetuate this relationship.

The Agta peasants, because of ethnic differences and different physical appearance, are easily recognizable. They are considered inferior by many Filipinos including those from the lowest stratum. Homesteading lowlanders have joined the ranks of medium-sized or family-size land owners. This has created a conflict situation. As Griffin (1991:205) has observed, "The goal of the poor, pioneering farmer, or the established 'simple technology' farmers, such as those distant from town centres, is to gain as much as possible in commodities and work from the Agta while yielding as little as possible." Some pioneering farmers encourage Agta to settle near their farms. They threaten and intimidate the Agta to prevent them from planting fields of their own. This forces the Agta to work for the homesteaders to obtain food. Or the homesteaders encourage Agta to clear land and plant, only to seize the land by theft, force, or legal maneuvering. Headland (1986:431–34; 575–92) has documented more than 30 cases of these disputes. Griffin (1991:206) notes that, besides threat and intimidation, lowlanders use three additional hooks to subordinate the Agta—tobacco, rice, and alcohol. Agta may attempt to establish traditional ahibay relationships with the pioneering lowlanders, but usually the newcomers want only wage laborers or they misunderstand the intent of the Agta. As a result the Agta are frequently exploited owing to misunderstanding or ill will or both (see Griffin 1991).

Headland (1988b; Headland and Headland 1997) has proposed a model of this conflict between peasant Agta and lowlanders to explain why foragers have difficulty becoming agriculturists. Borrowing an ecological term, he describes it as the "competitive exclusion principle." The principle states: "Local agriculturists do not want foragers to move into full-time farming. . . . They are dependent on their forager clients as suppliers of protein food, forest products and laborers, and will one way or another, try to block any attempt of foragers to become independent farmers." In areas of the Philippines where tribal groups such as the Agta are being incorporated into the rural structure as peasants, ethnicity becomes an added factor of class conflict.

The Three Phases of San Ildefonso Agta Culture Change: From Foragers to Peasants, 1950–1994

The Agta have undergone major socioeconomic change in the latter half of this century. Beginning with chapter 6, the book divides this acculturation process into three distinct historical periods for the San Ildefonso Peninsula. During the first period, called the Forager Phase, from 1950 to 1964, the Agta on the peninsula were still able to live their traditional hunter-gatherer lifestyle based on collecting and trading forest products with down-river farmers. The second period, the Transition Phase from 1965 to 1979, began when the Agta slowly moved from their foraging economy toward a livelihood based on wage labor as their forest resources declined. The third period, the Peasant Phase, from 1980 to 1994, describes their move toward peasantization as Filipino colonists moved onto their peninsula, loggers destroyed the forest, ecological degradation overcame the region, and a new "Acculturating population" emerged.

The main questions of this research are: What are the population dynamics of the San Ildefonso Agta before, during, and after their change from foragers to peasants? Has the population grown or decreased? The demographic explanation of population change is always a four-variable problem—fertility, mortality, in-migration, and out-migration. What was the level of each of these variables while the Agta were still foragers in the twentieth century, and how did the Agta change as they became peasants? Is the situation of the Agta as Filipino peasants better or worse than the period when they foraged in this remote region and exchanged with a lowlander population with whom they had a long-standing relationship? These questions will be examined in part IV. But prior to this, part III describes the methodology for the collection and verification of the data used in the examination.

Part III

Methodology

Chapter 4

Fieldwork and the Database

This chapter and the next deal with many small details comprising the demographic methodology. They explain the concepts, procedures, pragmatic judgments, fieldwork, and critical checking that went into the construction and verification of the quantitative database used for the analysis in parts IV and V.

DEFINING THE STUDY POPULATION

The purpose of the research was to study the population dynamics of the Agta. There are a number of theoretical and practical problems in determining the boundaries of a foraging population. The older conception that these societies are composed of relatively stable, clearly distinguished bands occupying fixed territories has been shown to be invalid. Among the Agta there is no overarching social structure that defines the group, merely a number of camps composed of kin-related nuclear families strung out along eastern Luzon with kin connections between some but not all of these camps. There are no fixed territorial boundaries. There is frequent migration of individuals among Agta camp groups, sometimes involving long distances. The 9,000 Agta in eastern Luzon may be conceived as a single group composed of Agta families with greater interaction with other Agta families who live in close proximity to them but with marital and other kin ties joining them to some distant camp groups. If lines were drawn on a map around the interaction of each camp group, there would be a series of overlapping circles up and down the coast and adjacent mountains. The unity among them comes from a similarity of customs and the speaking of several regional and mutually intelligible languages that evolved from a common Proto-Agta language a few thousand years ago.

Several methodological options were possible to surmount these problems of defining the study population. One was to enumerate a few key demographic variables and, on this basis, to select an existing mathematical model that would estimate the values for the other demographic variables. This method was rejected because it is not known whether the assumptions of these models fit this type of population, and in fact there are reasons to suspect that they do not. Any analysis based on them raises as many questions as it answers (see Harpending and Wansnider 1982:37; Wood 1987; Early and Peters 1990:128–31; Hill and Hurtado 1996:115–19).

It would have been preferable to define the study area as all of northern Aurora, the 700 square-kilometer home area of the Casiguran Agta population. Headland (1986:546 and 1989) made a demographic study of this area for the period between 1977 and 1984. But a time span of seven years is too short to understand the trend of the population dynamics. Headland's study was restricted to a linguistic population; it included 103 Casiguran Agta who were living outside of the Casiguran area in 1984, but it did not include the 75 Agta in-migrants to the area, since Casiguran Agta was not their first language. In addition, it was impossible to track the demographic history of many of these people. Therefore, it was decided to define an Agta population living in a specific geographical area in spite of the difficult problem of enumerating migration in and out of the area. The area would have to be one for which the necessary data could be obtained for a considerable time period. The unattainable ideal for such a study would include all 9,000 Agta living in eastern Luzon.

The Headlands arrived in the Philippines in 1962 as members of the Summer Institute of Linguistics (SIL), one of whose goals is to translate the Bible into the languages of indigenous groups throughout the world. They selected the Agta because of their previous anthropological interest in foraging cultures and established residence among the Agta on the San Ildefonso Peninsula. In spite of its remoteness, the Casiguran area had relative ease of contact by air and radio with the SIL logistics center at Bagabag in Nueva Viscaya.

An area that could fulfill both the methodological and practical requirements of data acquisition was the San Ildefonso Peninsula, a part of northern Aurora and within the municipality of Casiguran (see Map 2). This

109 square-kilometer (42 square-mile) peninsula is bounded by Casiguran Bay on the west and the Pacific Ocean on the east. On the north it is separated from the rest of the municipality of Casiguran by the Didumayan River, whose mouth is at approximately 122° 5' east longitude and 16° 10' north latitude. For analytical purposes, we have defined three distinct populations living on the Peninsula: (1) the Agta population, (2) the Filipino (lowlander) population, and (3) the Acculturating population (to be discussed later). This research analyzes the population dynamics of the Agta and Acculturating populations of the peninsula.

For all individuals in the database the specific years they lived in the geographical area were recorded from their demographic history of places and dates of birth, death, and migration. Special computer programs were written to compute and convert this information into annual sums of the demographic variables by age and sex.

FIELDWORK

The Headlands (Thomas is shown in the introductory photograph of this chapter) lived in the Philippines from 1962 to 1986. During this period, six years were spent on home leaves or in residence at the University of Hawaii. Five years were spent teaching or doing research in Manila and other parts of the Philippines. Six years were spent in the towns of Casiguran or Bagabag, Nueva Vizcaya, interviewing Agta, and seven years were spent living in camps with the Agta, mostly on the peninsula (Headland 1986:xxxii).

During their initial years the Headlands concentrated on the study of the language of Casiguran Agta. In 1974 they published an Agta-English dictionary (1974a) and grammar (1974b), and in 1977 they completed a translation of the New Testament in Agta. Following this they turned their attention to ecological and demographic questions. The Headlands gathered the following data sets during these years:

1. From 1962 to 1979, as the Headlands gradually came to know the people, they kept daily diaries in which they recorded an increasing number of births and deaths.

2. Between 1976 and 1979 a retrospective demographic survey was undertaken of Casiguran Agta in Casiguran and in neighboring municipalities. The survey included name, date and place of birth, migration, marriage and divorce, date and place of death, cause of death, genealogy, and fertility history. (For details, see Headland 1986:47–52.)

3. Between December 1982 and July 1984 dissertation research was undertaken among the Agta. It included a new demographic survey using the previous one as a basis, and studies of time allocation, swidden and wet rice agriculture, and types of food consumed.

The period from 1976 to 1984 was covered by the above research. But for the analysis here, a longer time depth was desired in order to understand any changes in demographic structure that may have accompanied the social changes discussed in the previous chapters. Both the critical methodology required to overcome underenumeration (to be discussed in the next chapter) and the enumeration of migration required the database be composed of records of individual people. The Headlands' records formed the core of the database. Additional data were needed on individuals who had died before 1976. Important information was provided by two previous investigations: The missionary anthropologist Morice Vanoverbergh (1937–38:140–47) visited the Casiguran area for two months and compiled a list of 476 Agta individuals by name from 59 families, including dead family members. Tomas Casala (1960), the government's superintendent of non-Christian tribes in Casiguran from 1960 to 1963, enumerated a census of 427 Agta, also by name, in the municipality of Casiguran. Thus the initial records of the database came from the fieldwork of the Headlands supplemented by the list of Vanoverbergh and the census of Casala.

COMPOSITION OF THE DATABASE

From these sources a computerized database with 185 variables was compiled employing the Paradox program. Below is a summary list of the 185 variables. All dates have three elements—year, month, and day. Marriage and migration can be repetitive events, so the variables describing them required up to six repetitions. Migration means not only in and out of the study population (external migration) but also between the three geographical subsectors on the peninsula (internal migration) to be discussed in chapter 10.

1. ID #
2–4. Names
5. Sex
6. Member of study population or not?
7. If so, what years?
8. Member of Acculturating population?
9. If so, what years?
10. Father's ID and name

11. Mother's ID and name
12. Mother's age at birth of this ID (derived from later analysis)
13. ID's year, month, day of birth
14. Birth date registered or estimated?
15. If registered, all or what part of date?
16. If estimated, range of possible error?
17. Place of birth
18. ID's year, month, day of death
19. Death date registered or estimated?
20. If registered, all or what part of date?
21. If estimated, range of possible error?
22. Place of death
23. Cause of death
24. Age at death
25. If a mother, quality of fertility history with regard to:
    a) all children
    b) all children who were members of study population
26. ID # and name of spouse
27. Year of marriage
28. Year of termination
29. Reason for termination
30. Year of in-migration
31. Reason for in-migration
32. Age at in-migration
33. In-migrated from ___
34. In-migrated to ___
35. Year for out-migration
36. Same for out-migration as 31 to 34 for in-migration

The information was coded for all the Agta for whom information existed, whether members of the San Ildefonso study population or not. Also included were lowlanders married to Agta and their offspring. This gave a total 3,563 records, of which only 1,073 had ever lived on the peninsula and only 857 since 1950, which was later determined as the starting point of the study. All the records were used to construct fertility histories and to help determine averages requiring data with exact dates.

DEFINITION AND DATES OF MIGRATION

Migration presented a problem because many Agta move frequently. It would be an impossible and unnecessary task to record all movements, as many are not significant demographically owing to their short duration (for example, to visit relatives in another area for a short period). Also, since much of the migration data would be retrospective, a stay in an area required a duration long enough so that a person or family came to be recognized as living in that area and could be remembered as such. For a nuclear family, an approximate three-year stay in an area was selected as the criterion, which meant that a family group must have stayed on or been away from the peninsula (or a subsector of the peninsula for internal migration) for approximately three years or more in order to be considered an in-migrant or out-migrant. However there were also instances where individuals moved about without their family groups. Some examples would be widows, widowers, divorced people, old people without dependent children, and young men looking for brides. These people would more quickly become identified as residents of an area than family groups. For individuals moving without family groups, an approximate one year of presence on or absence from the peninsula was selected as the criterion for counting the move as a "migration" rather than a "visit."

Quality of Age Enumeration

All calculations of ages in this study use the logical form: date of demographic event minus the date of birth. Therefore, the quality of the birth dates have an importance beyond the analysis of fertility. They are used for age-specific rates of all variables. Rates of this type are essential for understanding population dynamics.

There are only a few instances where exact dates are required. The most demanding case is the precise measurement of a birth interval, which requires two consecutive exact dates. This is an important measurement for evaluating the data. The average birth interval becomes the criterion for determining possible underenumeration in fertility histories and the need for further fieldwork.

For the calculation of the conventional demographic rates, age classifications of 5 and 10 years will be used. In addition, the 44 years of the time series are grouped into phases of 15 years. This grouping or collapsing of the data by individual years into these larger age and time segments diffuses much of the estimation error. Exact ages and exact birth dates are not required for most analytical purposes.

Registration

The term frequently used for exact dates is "registered" dates, although this does not mean the use of formal, legal birth registers. Here it means that the date of birth was noted down close to the time of the birth and is therefore known. Some are known because of a relationship to the memory calendar described below. Exactness can be of three degrees: (1) year, month, and day; (2) year and month; and (3) year only. Table 4.1 shows how many dates of birth of those who ever belonged to the study population were exact to any of the three degrees and how many were estimated. The table indicates that 32 percent of the births are exact at least as to year of birth. The dividing line in quality is around 1975. Before that date fewer than 10 percent were registered at least as to the year, but after 1974 approximately 85 percent were registered. The quality of enumeration of the dates of death shows the same dividing line in Table 4.2. Less than 8 percent are exact at least as to year in the years before 1975, while 81 percent are exact after 1974. This was expected, as 1975 was the year before the Headlands began their first systematic demographic survey.

Estimation

Most of the dates of births and deaths prior to 1975 required estimation. The estimation of age in terms of individual years is always a problem in nonliterate societies. The problems of attempting estimation of age from

Table 4.1. Birth Dates by Quality of Determination

Birth Year	All Births	Exact Date	% Exact	Estimated Date	Type of Exactness*		
					1	2	3
<1950	264	2	0.8%	263	2	—	—
1950–54	64	1	1.6%	63	—	—	1
1955–59	74	5	6.8%	69	2	—	3
1960–64	70	19	27.1%	51	5	5	9
1965–69	69	10	14.5%	59	4	1	5
1970–74	65	19	29.2%	46	6	2	11
1975–79	58	45	77.6%	13	4	19	22
1980–84	70	60	85.7%	10	19	19	22
1985–89	69	57	82.6%	12	8	18	31
1990–93	54	53	98.1%	1	24	16	13
Total	857	271	31.6%	586	74	80	117
<1950–74	606	56	9.2%	550	19	8	29
1975–93	251	215	85.7%	36	55	72	88

* 1—exact to year, month, and day
  2—exact to year and month
  3—exact to year

physical appearance alone are well known. People usually do not know their ages and often do not understand the question. The question comes from a cognitive framework foreign to groups such as the Agta. Yet the analysis of population dynamics requires age-specific rates for an understanding of the demographic variables. The following estimation procedures were used to calculate these age-specific rates.

Estimation When Day or Month Unknown

If the year and month were known, then the fifteenth of the month was used to estimate the day. If only the year was known, the midpoint of the year, July 1, was used as the estimate.

Age Calendar (Wheel, Computer)

For the surveys in 1976–79 and 1983–84, the Headlands employed a memory calendar of dated local events to estimate birth and death dates (Scanland 1976). Agta mothers were asked to relate the birth of their children to memorable events in local history for which a date could be established. If a person was born between events, the birth date was estimated

Table 4.2. Death Dates by Quality of Determination

Death Year	All Births	Exact Date	% Exact	Estimated Date	Type of Exact*		
					1	2	3
<1950	511	10	2.0%	501	—	—	10
1950–54	47	0	0	47	—	—	—
1955–59	35	1	2.9%	34	—	—	1
1960–64	44	5	11.4%	39	3	—	2
1965–69	55	20	36.4%	35	13	4	3
1970–74	37	16	43.2%	21	3	3	10
1975–79	35	28	80.0%	7	8	13	7
1980–84	59	53	89.8%	6	19	15	19
1985–89	83	58	69.9%	25	21	17	20
1990–93	44	40	90.9%	4	14	19	7
Total	950	231	23.2%	719	81	71	79
<1950–74	729	52	7.1%	677	19	7	26
1975–93	221	179	81.0%	42	62	64	53

* 1—exact to year, month, and day
2—exact to year and month
3—exact to year

by asking about a person's sibling order or physical characteristics at the time of the closest remembered event and interpolating.

The events used for the calendar were:

1912.	Arrival of American Army officer Wilfrid Turnbull to start reservation at Calabgan
1913.	150 Agta families move to reservation
1918.	Teacher Mrs. Emila Salsedo arrives at reservation
1932–40.	Teodorico Molina serves as local agent for Commission of National Integration
1933.	Agta men shoot sergeant
1934.	Three Agta men jailed for killing Muslim sailors at Dinapigui
1936.	Vanoverbergh survey
1942.	Reservation school closed
1942–45.	World War II. Japanese in Casiguran
1945.	Liberation. U.S. Navy stops in Casiguran
1956.	Agta man kills the town mayor
1959–63.	Tomas Casala is local agent for Commission of National Integration
1960.	Magnesium mine begins at Dinapigui
1961.	Many Agta work at mine
1962.	April. Headlands arrive
1968.	August. Famous earthquake
1970.	September. Typhoon Pitang destroys town
1972.	September. Martial law declared
1974.	May. Arrival of NPA guerrillas
1975.	February. Agta move to reservation
1976.	Headlands move to Calabgan
1979.	Headlands leave Calabgan
1980.	November. Typhoon Aring

THE SAN ILDEFONSO AGTA DATABASE

This chapter has described how the study population was defined and the Headlands did their fieldwork. The individual records of the database were compiled from this fieldwork, from Vanoverbergh's 1936 list of Agta names, and from Casala's 1960 Agta census. The chapter has described the estimation methods used to obtain the ages of those for whom there was not a registered birth date. The next chapter explains the methodology to check the quality of these data for underenumeration and errors.

Chapter 5

The Critical
Process

The preceding chapter described the fieldwork and methods that went into the construction of the San Ildefonso Agta database. Experience in this and a previous study (Early and Peters 1990) has shown that in spite of the long-term residence of researchers with the people of the study population, there are still problems of omissions and mistakes. People are omitted because they died very young or were not well known to the researchers. Other problems are double entry of the same person under another name, incorrect dates, and clerical mistakes while coding and entering data into the computer. Demographic data uncritically accepted, especially data collected by researchers in short-term residence (Early 1985), leave many unanswered questions about the validity of the analyses based on the data. In these types of populations there are no perfect databases. The goal is to minimize the problems so that their influence on the analysis is minimal. The methodological problem at this stage of the research is how to find the omissions and mistakes in the database and how to correct them. It is a laborious process. The omission of this process in many studies of the demography of anthropological populations results in mistrust of the completeness of many data sets and the results obtained from their analyses.

The critical process (1) subjects the original field data to an intense scrutiny for completeness and accuracy; (2) makes a log of the problems discovered; (3) with the log, returns to the original field site to investigate the problems; (4) revises the database using the results of the field trip; and (5) repeats the above steps until the database appears satisfactory.

THE CRITICAL TOOL: ANALYSIS OF FERTILITY HISTORIES

Any method that cross-checks the logical relationships within the data for internal consistency can be used for the intense scrutiny. The method used in this and a previous study (Early and Peters 1990) is to construct fertility histories for every mother in the database. Table 5.1 presents some examples of these histories and the information they contain. Practically every piece of information in a woman's fertility history must be logically consistent with other data in the same woman's history and usually other histories, and/or biological norms or social norms. (For example, if a history printout showed a woman giving birth six months after a previous birth, or a birth at age 55, a logical inconsistency—an error—was suspected.) Putting the data into the framework of fertility histories subjects them to a tight logical discipline and quickly indicates problems.

Table 5.1 shows the format used for the fertility histories. It would take a manual to detail all of its implications. Here only a few examples will be indicated. Information about the mother is contained on the left side, the father on the right, the child in the right center, and fertility information in the left center. Length of birth intervals (Spac for "spacing" in Table 5.1) is an important variable. Chapter 7 on reproduction will discuss the methodology for deriving the average length of various types of birth intervals from these fertility histories. The spacing for each interval of each history is compared with the values shown in Table 7.2. If the length of an interval appears excessive, it raises the question of a possible missing person in the history and therefore possible underenumeration of births and maybe of deaths. If the interval is too short, this raises the question of incorrect dates or the same person being entered twice in the database, usually with a different name. Realistic ages of the mother at first and last birth also provide criteria. Many mothers appear in two fertility histories: in that of her mother with her siblings, and again in the woman's own fertility history showing her with her children. Age coherence must carry through both of the histories. Any apparent problem concerning one individual means that the records of all persons connected to that individual by the histories must be checked for consistency.

Table 5.1 shows three examples of the preliminary fertility histories. In the first example, a mother called X is shown as having had only one child, but this becomes intelligible when the age of death of the mother is noted (in the column labeled "DAG"). The second history, of mother Y, appears to have underenumerated the children of this mother. However, the code 3 under "Rank" indicates that if there were more children, none of them

Table 5.1. Examples of Provisional Fertility Histories

			Mother													Fertility							
MID	Name	Age	B1	E1	B2	E2	B3	E3	DYR	DAG	Rank	BQ	TR	BER	BYR	MAGB	CBYR	Sib	Spac	BER	BQ	TR	DAG
58	X	53	62	65	—	—	—	—	65	24	1	1	2	2	40	21.0	61	1	—	1	1	3	1.5
59	Y	53	65	67	—	—	—	—	—	—	3	1	2	7	40	14.0	54	1	—	4	2	—	—
																19.0	59	2	5.0	3	2	2	
																23.0	63	3	4.0	3	2	2	
60	Z	49	56	57	60	94	—	—	—	—	1	1	2	2	44	15.0	59	1	—	1	1	3	19.5
																19.4	63	2	4.5	—	1	2	
																25.8	70	3	6.4	—	1	2	
																29.0	73	4	3.1	2	2	2	
																32.5	77	5	3.5	—	1	2	0.0
																36.1	80	6	3.6	—	1	2	
																37.6	82	7	1.5	—	1	1	
																41.4	85	8	3.8	—	1	2	

MID—mother's identification number

Age—in 1994 whether dead or alive; used to sort database by mother's age and to determine postmenopausal mothers

B1, E1—beginning and ending years for the first time the woman lived in the study population. Two additional periods are listed for women who moved into and/or out of the area later: B2, E2; B3, E3. (For example, in the last case, woman #60 entered the population in 1956 and left in 1957; she reentered again in 1960 and was still present at the end of the research period in 1994.

DYR—year of death. This cell is blank for woman #59, indicating she was still alive in 1994 and that she left the population in 1967 by out-migration, not by death.

DAG—age at death

Rank—Under rank there are two sets of codes for subjective evaluation by the field investigators about the quality of the sibling enumeration in a history. The first code is the quality of the enumeration of all the mother's live births. The second code asks the same question with regard to the enumeration of her children who ever lived in the research population. The code itself for both questions is a three-point scale: (1) morally certain that this history is complete with regard to the enumeration; (2) probably complete; (3) probably incomplete.

	Children											Father	
ID	Name	Sx	DYR	BV	DV	B1	E1	B2	E2	B3	E3	FID	Name
80	A	1	62	2	2	62	62	—	—	—	—	90	K
14	B	2	—	4		65	67	—	—	—	—	14	E
17	C	1	—	4		65	67	—	—	—	—	14	E
18	D	1	—	4	2	65	67	—	—	—	—	14	E
24	E	1	78	4		60	78	—	—	—	—	65	H
25	F	1	—	2		64	94	—	—	—	—	65	H
26	G	1	—	2		71	94	—	—	—	—	65	H
27	H	2	—	2		74	88	92	94	—	—	65	H
36	I	1	—	2		78	94	—	—	—	—	65	H
56	J	1	80	2	2	—	—	—	—	—	—	65	H
57	L	2	—	2		83	94	—	—	—	—	65	H
97	M	1	—	2		86	94	—	—	—	—	65	H

BQ—code indicating if birth date is registered (1) or estimated (2)

TR—if registered, code indicating if registered by day, month, and year (1), by month and year only (2), or year only (3).

BER—if BQ is estimated, number indicates possible range of error for birth year; a "2" indicates the person was born in the year shown in the next (BYR) column, plus or minus two years.

BYR—birth year

MAGB—mother's age at birth of child

CBYR—child's birth year

Sib—sibling order

Spac—length of birth interval (e.g., 4.5 means 4 1/2 years between this birth and that of the previous sibling

Sx—sex of child

BV—code for place of birth

DV—code for place of death

Later another column was added with a code for the type of birth interval, based on whether it included a fetal, infant, or child mortality.

lived in the study population, so it is not of concern. The mother may be too young—only 14—at the time of birth of her first child, because of the late age of first menstruation among Agta. An adjustment may be needed to either her birth date or that of the child, since the codes indicate that both were estimated. The third example, of mother Z, is a complete fertility history of a woman who lived until past menopause, and most of the birth dates were registered (1 under the "BQ" column). The 6.4-year spacing for the third child indicates there may be a missing birth that should be checked. This format for the fertility histories poses questions such as these, which need to be investigated during the critical process.

All the questionable cases were examined to see if they could be explained historically. Some cases of long intervals between births were because of widowhood—a husband died and his widow did not immediately remarry. In others the woman was known to have had difficulties in conceiving or bearing children. Some could be corrected from the Headlands' personal knowledge or additional records. For the remaining questionable or impossible cases, they were recorded in a log to be checked by additional fieldwork.

This usage of fertility histories includes the ranking method used in a number of previous anthropological studies (Rose 1960, Howell 1979). Siblings within a history must be ranked. For verification, information was also obtained on the relative ranking of individuals in different histories, although that information did not allow the construction of a master list for the entire population.

Some histories appeared deficient because there were few children. It was impossible to check some of these histories because the mother was dead or living elsewhere and no close relatives were in the Casiguran area. Some of these women were not well known to the Headlands, especially if they died or moved away before the Headlands arrived or before they ever met them. A list of these cases was drawn up. Using linear interpolation and average spacing norms, hypothetical children were added to these histories. It was also assumed that these hypothetical live births died in the first year of life. This seemed a reasonable assumption, since experience while checking other histories had shown this was a frequent reason for omissions in other birth histories that were better known. Still, the assumption involves some estimation error. Forty-seven hypothetical infants were created. But the estimated dates for only 37 of these placed them as being born in the study population while their parents resided there. In later interviews in 1992 and 1994, and in the Headlands' diaries, eight of them were verified as real infants and two as nonexistent. This left 27 hypothetical infants: 2 born during the 1930s, 15 during the 1940s, 7 during the 1950s, and 3 during the 1960s.

FIELDWORK—CONTINUATION OF THE CRITICAL PROCESS

In 1992 the Headlands returned to the San Ildefonso Peninsula for four months of interviewing the Agta and checking information. Their fieldwork was based on and focused by copies of the fertility histories and the log of problems previously formulated. The Headlands discussed each problem with the Agta. The database was also updated to 1992. Upon their return, the revised and updated information was entered into the database. New versions of the fertility histories were then generated and analyzed.

THE TEMPORAL STARTING POINT OF THE STUDY

A constant question during the critical evaluation of the database was how far back in time the demographic analysis could be validly taken. In order to detect variation in the demographic processes, the desire was for as long a time series as methodologically possible. The database included records of individuals who lived in the last quarter of the nineteenth century, but obviously the quality of the database deteriorated as one went backward in time. Demographic rates were calculated back to 1920. The rates for the 1920s and 1930s appeared unreliable. The question was whether to include the 1940s. Both the birth and death rates in the 1940s were noticeably lower than those of the succeeding 5- and 10-year periods. In the evaluation process, the highest number of hypothetical infants were estimated for the 1940s. The Headlands arrived in the Philippines in 1962. With their long residence in the area, mastery of the language and intimate knowledge of the people, reconstruction of the 1950s appeared feasible. But the data from the 1940s were raising too many questions. Therefore, 1950 was chosen as the temporal starting point. The data for the 1950s had not offered many problems either in the fieldwork or the evaluation process. As a result, of the 47 hypothetical infants originally added to the database, only the 10 assumed born after 1949 were used in the final analysis.

A FINAL FIELD TRIP—THE END OF THE CRITICAL PROCESS

Some problems remained or emerged after the 1992 field trip. The 1992 trip had highlighted the increasing number of Agta-lowlander marriages and the growth of these families. How to demographically conceptualize them remained undecided. Another trip would add two more years of data to the study. Headland and Early (shown in introductory photograph with survey helicopter) lived with the Agta on the peninsula for six weeks in 1994 to clarify these problems and to update the database. Janet Headland made a brief visit in May 1995. The final date for the data is December 31, 1993, which gives the size of the study population as of January 1, 1994.

THE CORRECTED SAN ILDEFONSO DATABASE

This chapter has presented the critical methodology used at every point possible to ensure the validity and completeness of the data. The critical process required the construction of more than 100 fertility histories, a lengthy period of correction and reconstruction, and two return field trips in 1992 and 1994 to obtain missing data or to correct existing data. Numerous additions and corrections were made. The experience of reconstructing a population with the methodology used here forces seasoned observers to look at the population in a way they had never done before. It brings up questions about matters they had never thought to observe. This study redefined the population and expanded on the previous demographic work of the Headlands (Headland 1989). The Headlands with their knowledge of the language and the people conscientiously carried out the fieldwork for the original study. The data were checked but in a nonsystematic fashion, as appears customary for most anthropological demographic studies. Nevertheless, when the data were subjected to the discipline of constructing fertility histories, some omissions were quickly discovered. Unless data have been subjected to this kind of critical scrutiny, a number of reported levels of fertility and mortality for anthropological populations appear to be too low because of underenumeration.

SOME NOTES ABOUT TERMINOLOGY

Before beginning the analysis, this section clarifies usage of some expressions which may be vague or unfamiliar to some readers.

Anthropological Demography and Anthropological Populations

These basic terms are vague because of historical developments. Anthropology is an evolving discipline whose boundaries are currently difficult to define. Here "anthropological populations" is used in an older sense. Historically cultural anthropologists studied small, nonliterate societies, and this is the sense in which the phrase is used here. Anthropological demography refers to the use of conventional demographic methods for the study of these small-scale societies. Sometimes anthropological demography needs to develop its own vocabulary and methods instead of blindly following demographic conventions appropriate to other types of populations.

Migration

The real estate of the planet is divided among nation states with international political boundaries. This type of cognitive map has resulted in the demographic convention of distinguishing internal migration, which is migration within a national boundary, from external or international migration, which crosses a national boundary. Those involved in the former are designated in-migrants and out-migrants, while those in the latter are designated as immigrants or emigrants. For the analysis of most anthropological populations, these distinctions based on international boundaries are culturally misplaced. Anthropological populations are not nation states. Many international boundaries were drawn with little knowledge or respect for traditional land usage by anthropological populations and are unknown or ignored by these populations.

The convention also fails to take into consideration the problem of definitions of a study population such as this one. This study involves migration in or out of the San Ildefonso Agta group, both physical migration, based on the geographical boundary, and cultural migration, based on the way of life. None of these migrations involve an international boundary. This study will distinguish three distinct areas (called subsectors in chapter 10) within the area of the study group with migration between them, that is, internal migration from one part of the peninsula to one of the other two sectors on the peninsula.

Sex Ratio

The sex ratio expresses the sexual balance or imbalance in a population. The conventional sex ratio is the number of males divided by the number of females and multiplied by 100. This is a male ratio, since it tells the number of males per 100 females in the population (Shryock and Siegel 1973:191). When the interest is the number of females, the expression becomes clumsy. For this reason this study will simply use a percentage form of males or females in the population. The conventional sex ratio can be calculated if needed.

Yearly Population Sizes

There are two values used in the analysis: the beginning of the year (the same as the end of the previous year) and the midyear value. From the coding for each individual record in the database, a computer program determined where a person was residing at the beginning of each year. These were summed to give the size of a group's population at the beginning of the year. Midyear values were used for all crude rates. These values were derived by linear interpolation from the beginning-of-the-year populations for two consecutive years. Because of the differences between these two types of annual population values, demographic rates for the same year or periods of years can differ slightly, depending on which population value is used in the calculation.

Rates per 1,000 Population

Crude rates by definition are expressed as per 1,000 population, since conventional demography has its roots in the analysis of large groups. Consequently most demographers are accustomed to thinking in these terms. However, many anthropological populations do not have populations that reach 1,000 people. It could be argued that anthropological demography should use a convention of per 100 population or less. But for purposes of comparison, it is more convenient to continue the convention and understand the reason for it.

An exception to this is the discussion of rates of natural or total increase. Many nondemographers, especially economists, prefer to express this as per 100 population (percentage) rather than the per 1,000 form. Both forms will be used here to express rates of population increase.

Ambiguity of the Term *Fertility* in Foraging Studies

In this study *fertility* will be used in the conventional demographic sense. This is a live birth defined by an infant showing vital signs upon its expulsion from the birth canal, no matter how quickly death may follow. If there are no vital signs, then a still birth has occurred, but demographically this is a fetal death with no demographic birth having taken place. An infanticide is both a birth and a death.

There is another meaning used for *fertility* in the literature about foragers, and data collected with this definition cannot be directly compared with data using the conventional definition of fertility. This alternative usage is valid but restricted. It is noted here only to prevent confusion for those who may be unaware of it. Under the field conditions of foraging societies it can be difficult to make the distinctions demanded by the conventional definition. Carr-Saunders, who was important in initiating the discussion of forager fertility in the 1920s, used an alternative concept. It is called effective fertility, although "effective" is sometimes dropped, creating the ambiguity with conventional fertility.

Because of the difficulty of enumerating infants who died very young, effective fertility simply removes them from the concept of fertility. The most frequent usage is to remove infanticides. In this usage, effective fertility is defined as the number of children a woman intends to raise. This usage tends to be more common in the older literature such as Carr-Saunders (1922:52); Krzywicki (1934:136, 217); Birdsell (1968:236); Dumond (1975:720n.34); Masnick and Katz (1976:39). If these authors contend that foragers have high rates of infanticide, and most do, their position is that effective fertility is low, but this appears to imply that demographic fertility is high.

Owing to the difficulties of interviewing, another version of effective fertility was used by Harpending (1976:156) in his study of the !Kung. His fertility counts included only infants who survived the first three days of life. This was based on the fact that the !Kung do not name the infant until this time and only then consider it a person. Because of this definition, Harpending confined the use of his results to internal comparisons.

Summary

This chapter has examined the critical process. It puts the demographic data collected under the difficult conditions of anthropological societies into a logical framework to check for underenumeration and correctness of dates for those enumerated. The critical process usually generates an agenda for further fieldwork. This chapter has also looked at the usage and assumptions of some terms. With the completion of the construction and verification of the database, parts IV and V employ it in the analysis of the demographic variables.

Part IV

The Demographic Structure

Chapter 6

Overview of
the Population,
1950–1994

The San Ildefonso Agta population grew from 178 at the beginning of 1950 to 231 at the beginning of 1994, an increase of 53 people in 44 years. Figure 6.1 graphs the yearly populations and the annual rates of increase. Table 6.1 shows these same populations and rates for 5- and 10-year periods. Since most of the yearly dates before 1975 are estimated (Tables 4.1 and 4.2), the 5- and 10-year rates for that period are more accurate than the annual rates. Overall the population grew at an annual rate of 0.62 percent (6.2 per 1,000 population per year), a low rate of increase compared with other populations.

RATE OF GROWTH AND ITS COMPONENTS

Although the overall rate of growth for the 44-year period is low, the population was volatile, as indicated by the range of growth rates. Table 6.1 shows that the highest growth rate for a 5-year period was 3.7 percent and the largest rate of decrease was -6.0 percent. There were two other 5-year periods when the increase was over 3 percent and another 5-year period when the rate of decrease was -1.4 percent. If these rates are collapsed into 10-year categories for 1950 to 1990, three out of the four decades have positive growth. The variation in these rates suggests some diversity of population dynamics over the 44-year period. This diversity would be the result of the interaction of the four demographic factors and/or the age-sex structure of the population.

The Demographic Factors of Population Change

Table 6.2 presents the absolute and relative values of the four components of population growth: births, deaths, in-migration, and out-migration. Group fertility is very high, as expressed by a crude birth rate of 53.3. Mortality is extremely high, as indicated by the 42.7 crude death rate. The difference between these two gives a 10.7 crude rate of natural increase (1.1 percent), a low rate compared with those of many agricultural societies but similar to those of industrial societies. In-migration has a high crude rate of 56.9, while out-migration is even higher with a crude rate of 61.3. These high rates indicate the mobility characteristic of foraging groups even when they become semisedentary. Their interaction resulted in a -4.5 (-0.45 percent) crude rate of net migration. This loss subtracted from the rate of natural increase means that the population grew by a low 6.2 (0.62 percent) crude rate of total increase during the 44-year period.

Table 6.2 and the tables in the following chapters give the absolute number of cases (N) of the demographic events (fertility, mortality, migration)

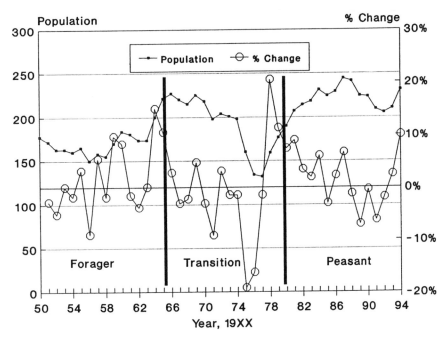

Figure 6.1. Size and annual percentage change of Agta population, 1950–94.

along with the rates. These rates are based on complete enumerations of the demographic variable or a reasonable approximation. They are not samples of the population, so there is no sampling error in this respect. Other researchers may wish to view these rates as samples of whatever construct with which they may be working. The N allows them to calculate the sample error in that type of situation.

The Age-Sex Structure of the Population

Changes in the four components of population growth can alter not only the rate of change of a population but also its age-sex structure. This phenomenon takes place when the components do not equally affect both sexes and all ages of the life cycle. However, Table 6.2 shows small differences in the percentages of the age-sex structure between 1950 and 1994, and the

Table 6.1. Size of San Ildefonso Agta Population and Average Annual Rates of Increase, 1950–94

Year (Jan. 1)	Population	Average Annual Increase (%)		
		5 Years	10 Years	Total
1950	178			
		-1.4		
1955	165		0.4	
		2.2		
1960	184			
		3.7		
1965	221		1.7	
		-0.3		
1970	218			
		-6.0		
1975	160		-1.4	0.6**
		3.5		
1980	190			
		3.3		
1985	224		1.6	
		-0.1		
1990	223			
		0.9		
1994*	231			

* 1990–94 is a four-year interval.
** Over the 44-year period, the population grew at an annual rate of 0.62% (6.2 per 1,000 population per year).

Table 6.2. Overview of San Ildefonso Agta Demographic Structure and Change, 1950–94

	1950	1950–94	1994	Change
Population	178		231	+53
Crude total increase		6.2 (53)		
Crude natural increase		10.7 (91)		
Crude birth		53.3 (455)		
Crude death		42.7 (364)		
Crude net migration		-4.5 (-38)		
Crude in-migration		56.9 (485)		
Crude out-migration		61.3 (523)		
Age-sex structure				
Male	88		117	+29
Female	90		114	+24
% Male-female	49.2%-50.8%	50.6%-49.4%	1.4%	
r male		+0.68%		
r female		+0.54%		
	%		%	%d
% Male age 0	17.5		17.4	-.1
% Female age 0	17.5		20.9	+3.4
% Male age 15	15.8		15.2	-.6
% Female age 15	15.3		10.4	-4.9
% Male age 30	9.0		10.4	+1.4
% Female age 30	11.3		9.6	-1.7
% Male age 45+	6.8		7.8	+1.0
% Female age 45+	6.8		8.3	+1.5
Total	100.0		100.0	0

r = the percentage annual growth rate
d = difference (the difference between the 1950 column and the 1994 column)

sex ratio is balanced. Therefore, any changes in the rate of increase (decrease) of the population are mostly due to the four demographic variables.

LEVELS OF POPULATION ANALYSIS AND THEIR EXPRESSION BY DEMOGRAPHIC INDICES

The next four chapters will examine each of the four demographic variables and their interactions. The results of the analysis are presented in demographic tables. To assist readers who may not be fully familiar with demographic terminology, this paragraph attempts a brief explanation so that the tables are more intelligible. Demography has established a set of quantitative conventions that include three differing but related levels of analysis. It is important to distinguish among these levels, because they do not necessarily move in tandem and may move in opposite directions. In other words, one level cannot be automatically projected from another. These levels are: (1) the total study population; (2) a part of this population, usually an age-sex segment; and (3) the average individual either in the total population or in a part of it. Different kinds of questions may emphasize one of the three levels. If the focus is the total population, crude rates of the demographic variables give the information. If the focus is demographic subgroups, the population is broken down into separate age-sex groups, and age-specific rates express the demographic levels for them. If the focus is an average individual, fertility analysis uses the average woman in the 15–44 female age group. Mortality analysis uses the average individual in the study population at the time of his or her birth. The reason the rates for the three levels do not necessarily move in tandem is that the relative sizes of the age-sex subgroups in a population may change. For example, women in the 15–44 age group may increase their fertility, but if their proportion of the total population decreases, the group fertility rate may decline. To properly understand the demographic dynamics of a group, rates for all three levels are needed.

PHASES OF SOCIAL CHANGE

The first three chapters have established that in the 44 years between 1950 and 1994, the San Ildefonso Agta have undergone a period of social change. For most purposes annual data for a 44-year period are too discriminate for a demographic analysis of these changes. The annual data need to be grouped together by a classification that distinguishes the phases of social change. The historical overview of the first three chapters suggests a threefold classification. The first is the Forager Phase. The third is the Peasant Phase, in which the Agta are incorporated into the dynamics of rural Filipino society. The second, in-between stage is the Transition Phase, as the change has been a gradual process rather than an instantaneous phenomenon.

The next task is an approximation process, in which specific years are designated as the beginning and end of each phase. This step allows the annual demographic data to be assigned to their proper phase. The beginning of the Forager Phase is the starting point of the study, 1950. The historical and ethnographic data clearly show that the Agta were still living primarily as hunters and gatherers at that time. The year 1964 marks the end of this first phase. Although the preceding cultural accounts have mentioned the 1960s as the time in which Casiguran lost its semi-isolation, it is evident that the process was just beginning in the early 1960s. The Headlands arrived in 1962, when the changes were beginning and had not yet had their full impact. The Headlands' medical program among the Agta (to be discussed in chapter 8) took several years to develop. Therefore, 1964 was chosen as an approximation of the end of the Forager Phase.

The beginning of the Transition Phase is 1965. This is the period in which the semi-isolation of Casiguran was definitively broken and the gradual incorporation of the Agta into rural Filipino society began. There was continued legal logging until 1978. The largest in-migration of homesteaders took place in this second phase, although it was interrupted by the NPA guerrilla warfare in the mid 1970s. The participation of the Agta in the guerrilla war was a significant experience of transition. Finally the opening of the road to Baler and the nation took place in 1977. Given a few years for its effects to be felt, 1979 marks the close of the transition period.

The Peasant Phase began in 1980. Agta and lowlanders were now interspersed with each other throughout the disappearing forest. The Agta have become part of rural Filipino society at the lower end of the stratification ladder. The ratio of lowlander to Agta continued to increase, and intermarriage between the two groups rose significantly (to be discussed in chapter 11). The final date in the database, January 1, 1994, is the end of the Peasant Phase for the purposes of this study. These three phases will be frequently used to classify the demographic indices in the tables.

Chapter 7

Reproduction

There were 455 live births in the San Ildefonso Agta population during the 44-year period: 216 males, 200 females, and 39 of unknown sex. To understand Agta reproduction, this chapter examines the mating pattern, the setting in which births take place, the female reproductive period, and the demographic levels of Agta fertility.

THE MATING PATTERN

Marriage is almost universal among the Agta. Mate selection is limited by the prohibition against marriage between relatives. Knowledge of the Agta kinship system is necessary to understand the implications of this prohibition.

Agta Kinship System

Anthropologists have classified a number of logical systems by which people relate themselves to one another as kin. In most industrial societies this system is known as the Eskimo system, the name taken from one of the groups who utilize it. This is also the system used by the Agta. Its basic unit is the nuclear family. There are no larger kinship units such as lineages or clans. It is a bilateral system with kin of both parents of equal importance. Marriage creates additional kin. There are some differences between the Eskimo system as employed by the Agta and that used in the United States: (1) Siblings are distinguished by a different kin term as to whether they are younger or older than the individual of reference; (2) kin terms do not distinguish siblings by sex; (3) sibling terms may be extended to cousins who are very close to the individual of reference. (For a formal analysis of the system, see Headland 1987b.) Therefore, each Agta child has his or her set of personal kindred shared only with his or her full siblings.

Marriage Prohibitions

Agta are forbidden to marry any person whom they already call by a consanguineal or affinal kinship term. Since an Agta's kin includes all conceptually recognized descendants of their eight grandparents, the marriage prohibition excludes a sizable number of people in this small population. It will usually exclude every person in the individual's own camp group as well as most of those in nearby groups. In this society a high percentage of people with whom an Agta daily interacts would usually be excluded as a possible marriage partner by this incest rule.

Mate Selection

When it comes time for a young man to marry, his parents select the prospective bride and enter into negotiations with her family. Marriage is essentially the union of two family groups. The concept prevalent in industrial societies of a contract between two individuals is unknown. The boy's parents may seek the services of a marriage petitioner to help them with the request. This is usually a lowland farmer who is a regular trading partner (ahibay) of the boy's parents. The petitioning party is made up of the boy's parents, the parents' siblings, the petitioner, and sometimes the boy.

The Petition

The petition may consist of up to three formal visits with the girl's parents and her other relatives. During these visits, the petitioner and the boy's family request the girl by delivering stylized speeches rich in metaphor. The meanings of these metaphors are marked by brackets in the following example (Headland 1978:130).

> (Boy's spokesman): It will be good if you accept us, because it is the custom that when there is a "flower" [the girl] there are also large "flies" [suitors] coming around. . . . For example, when it comes to "land" [the girl], whenever we cut a trail to begin to make a "swidden" [arrange a marriage], we know there is no previous claimer to that "land." But if you can show us that there is someone else claiming

this "land," then we will not proceed. But if you show us that there is no other claimer, then let us go ahead and cut a "swidden."

If the petitioning is successful, the girl's parents eventually assent. Subsequent petition visits (usually two) discuss the length of the bride service by the boy and the gifts that are to comprise the bride price. All the boy's kin contribute to the bride price.

Bride Service

After the second or third petition, the young man begins a period of bride service by living with the girl's family and supporting them for a previously agreed upon period. This may be from three months to a year. If all goes well during this trial period, the pair may begin to sleep together. At the same time the prospective union may be called off at any time by either family.

The Wedding

If the period of bride service has been performed satisfactorily, a wedding (kasal) may be held. This consists of a feast celebrated by the two family groups with plenty of nipa-palm wine or coconut wine. During the celebration the gifts comprising the bride price are distributed to the girl's kindred. Usually these gifts include kettles, cloth, rice wine, bolos, and dried meat.

Alternative Forms

The previous description gives the ideal process of bringing together a reproductive couple. However, this outline is not always followed in practice. In a 1977 survey the Headlands found that only 44 percent (N=63) of the married males had performed bride service. Sometimes a boy and a girl simply run away for a few days to the camp of a friend or distant kinsman. They then return to the camp of either set of parents and live as a married couple. Individuals who are widowers or widows simply begin to live together as a married couple. As will be seen in the section on mortality, widowhood is a frequent occurrence among the Agta.

Divorce

Because of trial marriage and the difficulty of determining an exact time when Agta marriages begin, divorce here has the broad meaning of the termination of a union whether considered a trial marriage or a marriage.

It is unusual for couples with dependent children to divorce. Headland encountered only two such cases in his years among the Agta. In 1977, 18 percent of the ever-married adults (N=127) had been divorced at least once. Most of these took place during the period of bride service. Little reproductive time is lost in this population because of divorce.

Giving Birth among the Agta

Preparations

Preparations for a birth begin around the eighth month of the pregnancy. If the mother of the pregnant woman is alive, the pregnant woman with her husband and children will move to the mother's house for the birth and remain there for several weeks afterward. Thatch siding will be placed on the side of the house to protect against wind and rain. The couple buy new cloth or beg used cloth in town to wrap the newborn. Incense made from the resin of the almaciga tree (*Agathis philippinensis*) is burnt under the house at the place over which the mother will sit while giving birth. This custom is to protect the fetus from malignant spirits who may attempt to eat it.

Labor and Birth

When an Agta woman goes into labor, everyone in the camp including children may crowd into her house. The mother of the woman takes charge of the situation, giving advice and orders. An adult may hike downriver to summon the assistance of a lowlander midwife, although sometimes an Agta will perform this service. The person summoned is usually a lowlander trading partner of the woman's family.

The woman giving birth sits on the floor of the house and leans back against her husband, who holds her in a sitting position with his arms around her. At all times she is covered from the waist down with a large cloth or cotton blanket. Her hands may be beneath the blanket as she attempts to pull the baby out of the birth canal. The woman's mother and the midwife face her with their hands also below the blanket as they assist in the expulsion of the newborn. They work by the sense of touch without looking under the blanket. Many of the onlookers yell instructions to the mother and midwife about pushing and pulling or not to do so. Some will push on or knead the woman's belly. The Headlands believe some of this activity is excessive, having observed flesh wounds on some stillborns. The woman's face, forehead, and stomach are lightly rubbed with leaves of the pepper vine (*Piper betle*) and sometimes with crushed petals of marigolds (*Tagetes* sp.).

After the Birth

The infant is kept under the blanket for several minutes. When the infant cries, the group relaxes, relieved to know of the successful birth. The blanket is pulled back, and the infant is examined to determine its sex. If the newborn is dead, people are upset and may begin blaming mother or father or midwife or all three. The umbilical cord is cut with the sharp edge of split bamboo, although bolos, razor blades, or arrowheads are sometimes employed. Frequently these instruments are dirty. A small handful of ashes from the hearth is placed on the infant's navel to cover the piece of attached umbilical cord, which is covered with a strip of cloth wrapped around the body; sometimes the ashes of this poultice become cold and soiled by dogs that frequently lie on the warm hearth. Further kneading of the woman's stomach expedites the expulsion of the placenta. It is wrapped in cloth and buried in the ground close to or under the house. The newborn is wiped off with leaves and dry rags and then wrapped in cloth.

The First Month

Infants may be sponge bathed with warm water two or three weeks after birth. During the first month, fathers daily wash the swaddling of the infant as well as the clothes and blanket of the mother. Care is taken to protect infants from dampness and wind. During the first month incense is burnt and charms (especially garlic) are hung around the necks of the infant and mother, protecting them against harmful spirits. If an infant starts to cry at night, the parents and grandparents will immediately begin to hoot loudly to drown out the sound of the crying and thereby prevent any nearby spirits from hearing the infant and harming it. If the birth has taken place at the camp of the mother's parents, the family with its new addition returns to its own camp after several weeks. (As mentioned in the methodological section, the return to her parents' house for a birth is considered a visit, not a migration. All births were listed for the couple's camp of residence at that time, not the camp of the mother's parents, unless the mother-to-be had been residing there matrilocally throughout the pregnancy.)

THE FEMALE REPRODUCTIVE PERIOD

This section investigates the age of a woman as she passes through the various events of the reproductive period. These ages are summarized in Table 7.1. (While the demographic rates are based on complete enumeration, the indices in this section are based on samples.)

Menarche

Menarche is not marked by a ceremony among the Agta. Therefore it is difficult for ethnographers to observe the first occurrence and obtain exact data. Based on their conversations with the Agta, the Headlands agree with the estimate of 17.1 years obtained by the Griffins (Goodman et al. 1985:171) for the Cagayan Agta. Compared with some other populations, this is an older age for menarche to take place. Chapter 8 discusses the health of the Agta and some possible reasons for the late menarche.

Age at First Marriage

Cohabitation may take place before menarche, but it usually occurs afterward. In the database for females married since 1950 and for which there is a basis for a reasonable estimate of date of marriage, the mean age is 19.3 years, the median is 18.7, and the mode is 18 (N=69). Several women who first married in their late 20s or 30s increased the value of the mean.

Age at First Birth

The mean average age of mothers at their first birth for infants born between 1950 and 1993 is 20.4 years, while both the median and mode are 19.0 years (N=93).

Length of Birth Intervals

Table 7.2 shows the average length of the birth intervals, that is, the time between two consecutive live births. These averages are used not only here

Table 7.1. Ages of the Agta Female Reproductive Cycle

Event	Age	Years	N
Menarche	17.1		74
First marriage	18.7		69
First birth	19.5		93
Average birth interval		2.82	160
Second birth	22		estimated
Third birth	25		"
Fourth birth	28		"
Fifth birth	31		"
Sixth birth	34		"
Seventh birth	37		"
Eighth birth	40		"

Source: Menarche data from Goodman et al. 1985

in the analysis of fertility but also in the analysis of the fertility histories during the critical process discussed in chapter 5. To avoid error because of estimation of dates, only intervals are used where the birth dates for two consecutive children were known to be accurate at least as to year and month. In the total database there are 160 intervals that fulfilled these requirements. Their average length is 2.82 years. The Agta do not employ any conscious form of birth control or birth spacing. Children are highly desired. The contraceptive effect of lactation is not realized by Agta women. Occasionally there will be talk of someone who knows about a contraceptive liquid. But there is no widespread use of it, and the Agta themselves question its effectiveness.

There is no induced abortion of which the researchers are aware, even though this question was pursued with numerous Agta. Vanoverbergh (1937–38:129) mentions a postpartum abstinence of six months, but inquiries indicated that this reported custom is not observed. Inquiry found that there was no concern for birth spacing because of any constraints caused by camp mobility.

Types of Birth Intervals

Table 7.2 classifies the intervals into four types based on whether a fetal or child mortality occurred during the interval. In eliciting data about fertility histories, as much information as possible regarding stillbirths and late miscarriages was also collected. These fetal deaths lengthen an interval between two live births. The average length of this type of interval was 3.1 years.

An Agta child is nursed until the next pregnancy. The stimulation of the breast by the sucking of the nursling secretes the hormone prolactin in the mother's system. This hormone, as long as it is secreted in sufficient and timely quantities, has a contraceptive effect. Agta babies are nursed upon demand and sleep with their mothers so that the nursing takes place frequently. The contraceptive effect diminishes as the child grows older and receives supplementary food. With the appearance of the child's molar at about two and a quarter years, the nursing often continues but with less intensity, and it usually loses its contraceptive effect. If a young nursling should die, the lactation ceases and the possibility of another pregnancy arises. Two types of intervals have been distinguished on this basis. The shorter interval is one in which the nursling dies in the first month of its life (a neonatal mortality). Where such an infant death occurs, the average length of the interval is 2.3 years. Where the nursling becomes a child death be-

Table 7.2. Length of Birth Intervals

Type Interval Includes	Avg. Length Years	N	%
Fetal mortality	3.1	5	3.1
No early mortality	3.0	102	63.8
Child mortality	2.4	17	22.5
Neonatal mortality	2.3	36	10.6
Total of all intervals	2.8	160	100

tween 1 month and 2.25 years, the average length of the birth interval is 2.4 years.

Finally there are regular intervals, that is, intervals during which no recorded fetal, infant, or child (as used here) death occurred. The average length of the regular interval is 3.0 years. Although the number of cases is rather small for some types of intervals, the resulting averages of the types of intervals fall in the hypothesized order of magnitude. The intervals containing infant and child deaths are the shortest while the intervals with fetal deaths are the longest. The regular intervals are intermediate. One-third of the intervals are shortened by a mortality of a previous sibling who does not reach 2.25 years. This topic will be examined in the following chapter on mortality.

Subsequent Births

Using 19.5 years as the age of a woman at first birth and the spacing factor of 2.8 years, Table 7.1 estimates the typical age of the mother at the birth of the later children. Table 7.3 shows that the average Agta woman who remains alive for the full reproductive period will have seven children. Table 7.1 shows the last birth will take place during the woman's late 30s. Those with higher than average fertility will have another child in the early 40s. This conclusion agrees with an analysis of 46 fertility histories of mothers who survived the reproductive period. It showed the average age at the birth of the last child was 39 years.

Menopause

This research has no quantitative data on the age at menopause. Observation and conversation with Agta indicate it takes place during the early

40s, an approximation that agrees with the estimate of 43.9 years by the Griffins for the Cagayan Agta (Goodman et al. 1985:175).

The Reproductive Pattern: Overall Levels of Fertility

Table 7.3 presents the Agta fertility rates for the three levels of analysis discussed in chapter 5. The average female Agta who remains alive for the full period of reproduction has seven live births, as indicated by the total fertility rate. The age-specific rates give a picture of how fertility is distributed over the years of the female reproductive period. The rate for ages 15 to 19 years appears low compared with the other ages. This is due to the late menarche and consequent late age at first birth by Agta women. Group fertility expressed by the crude birth rate has a very high level of 53.2. Table 7.3 shows the Agta to be a high fertility population.

The Phases

Has there been any significant change in the level of fertility through the three phases discussed at the end of chapter 6? Table 7.3 provides information on the fertility variables for each of the three phases. The main questions are: (1) Why does fertility appear to decrease in the Transition Phase? (2) Why does fertility recover in the Peasant Phase and reach a total fertility level of 7.6 live births in that third phase?

Decrease in the Transition Phase

There was some decline of group fertility in the Transition Phase (1965–79), as shown by the 18 percent decline of the crude birth rate to 47.7. Some of this is due to the 2.6 percent decline in the proportion of the female reproductive population in the total population. But there was also a decline of total fertility from 7 to 6.5. An inspection of the age-specific rates indicates the decline took place in the age 15–19 category and in the above age 35 category. There may be some estimation error of ages in the 15–19 age category because the rate of the following 20–24 age category is extremely high. The decline seems mainly in the above age 35 category. There is no clear-cut answer. One possibility is that this was a period of unrest with the arrival of guerrilla warfare in 1973, the NPA, and the forced relocation of many Agta to the towns and the Calabgan reservation by the army in 1975, and the Agta's subsequent return to their home areas on the San Ildefonso Peninsula a few years later. This phase of transition also saw the depletion of the game on which the Agta had previously depended. Many Agta were still trying to make new subsistence arrangements to sub-

Table 7.3. Agta Fertility Rates

Level of Analysis	Rate	Age	Total	Phase*		
				F	T	P
Individual female	Total fertility		7.0	7.0	6.5	7.6
Reproductive	Age-specific	15	126	121	107	161
population		20	299	265	350	283
		25	344	370	315	341
		30	282	245	269	320
		35	213	208	137	294
		40	144	193	115	130
	% of total pop.	15–44	22.3	24.8	22.2	20.3
Study population	Crude birth		53.3	58.0	47.7	54.7
Number of births						
	Total		455	149	137	169
	Male		216	61	71	84
	Female		200	55	60	85
	Unknown		39	33	6	—
	% females of known sex		48.1	47.4	45.8	50.3

* F = Forager Phase; T = Transition Phase; P = Peasant Phase

stitute for the loss of previous resources. All these factors with the physical and mental stress involved may have been more disruptive for the 35 and older population and mitigated their usual level of fertility.

The Peasant Phase

The Peasant Phase (1980–94) has the highest rate of total fertility, an average of 7.6 live births for the mother who remained alive for the entire reproductive period. An inspection of the age-specific rates indicates that most of the increase was among women in their 30s. The reason is unknown. In spite of the highest total fertility for the three phases, the crude birth rate for this period was slightly less than the Forager Phase (1950–64). One reason for this decrease was the 4.5-percent decline of the female population of reproductive age in the total population.

SUMMARY

This chapter has examined the reproductive pattern of the San Ildefonso Agta. Age at menarche, at first marriage, and at the birth of the first child are relatively late compared with some other anthropological populations. Since there is no conscious effort to space or limit births and the contraceptive effect of lactation is shortened by high infant and child mortality, the average birth interval is short, about 2.8 years. The average woman will have around seven live births, four in her 20s and three in her 30s. The few women with higher parities will bear children in their 40s.

The analysis has shown that the reproductive period for the average Agta female lasts 24 years (43 years at menopause minus 19 years at first marriage). This conclusion assumes she survives the reproductive period. Seven pregnancies of 9 months is 5.25 years for the entire period. Nursing these infants for two and a quarter years means 15.8 years for lactation over the same period (although the next chapter will show that a number of infants die before reaching two and a quarter years). Combined, these figures mean the mother is pregnant or nursing 21 of the 24 years, or 87.5 percent of her reproductive span.

Chapter 8

Mortality

From 1950 to 1993 there were 364 deaths among the San Ildefonso Agta, 163 males, 161 females, and 40 infants of unknown sex. This chapter looks at the Agta view of death, the customs surrounding the time of death, the extent of mortality, and some of the reasons for it.

DEATH AMONG THE AGTA

When people think a relative may be near death, they will send someone to fetch a local healer, who may be an Agta shaman, a lowlander herbalist-diviner, and/or the American missionary. They will also send someone to summon a close consanguineous relative of the dying person if none is present. This precaution is taken so that the primary kin of the dying person will not blame the surviving widow(er) or anyone else in the camp of improper care of their dying kinsman, especially if he or she is young or the death is sudden or unexpected.

Agta activity at an approaching death is partly guided by their worldview. Headland (1987a) notes:

> The Agta believe in a single high god and in a large number of super-natural spirit beings that inhabit their surrounding natural environment. Depending on the class of spirit, these various beings live in trees, underground, on rocky headlands or in caves.

There are two general classes of spirit beings in the Agta worldview: *hayup* ("creature") and *bélet* or *anito* ("ghost"). The latter are always malignant. Ghosts are wandering disembodied souls of deceased humans. The ghosts of recently deceased adult relatives are especially feared, as they are prone to return to the abode of their family during the night, causing sickness and death.

There are several varieties of *hayup* creatures. Although these are non-human, they are bipedal and may appear in human form. Most varieties of hayup beings are malignant; others are neutral, and a few can be called upon for help in curing disease.

When the relatives think that death is imminent, shouting and excited activity begin. Some may pick and crush leaves or flowers of aromatic plants and place them directly beneath the nostrils of the dying person. Others with shamanistic powers will spit on the top of the head or belly of the dying person while demanding in a very loud voice that the malignant spirits of the forest depart from the house (Headland 1987a; see Figure 8.1). Others, especially a parent or spouse of the dying person, wail and cry out hysterically. Some shout to the dying person, "Don't die! You are dying now! Don't do that!" Once the person has expired, people begin crying, calling to people in nearby houses or shouting at the deceased, "Don't leave us! Don't leave us!"

Upon death, the body is laid out on a mat in the house, sponge-bathed if it is dirty with blood, vomit, or feces, and covered with a blanket. As word of the death spreads to neighboring Agta camps and lowlander farms, people will travel up to two hours to the house of the deceased for the wake. They enter the house and draw back the blanket to look at the face of the deceased. The closest relatives take turns sitting on the mat with the body throughout the day and night of the wake. If roundworms emerge from the mouth or nostrils of a corpse (often the case with children), they are removed with a leaf and dropped into the fire. Periodically the women (occasionally the men also) will break into loud weeping and wailing. During the night most adults stay awake drinking coffee and telling stories, many of which have little to do with the deceased. Most people seem to enjoy the wake, as it gives them a chance to socialize and to show respect to the grieving family.

The next morning some men will go to the place of burial and dig a grave about a meter deep with their bare hands, bolos, or a shovel borrowed from a lowlander. Agta do not have special graveyards, although

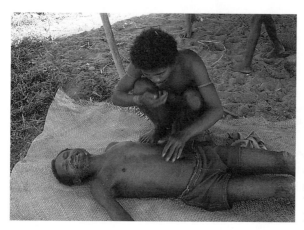

Figure 8.1. An Agta shaman spits betel quid on his hand and smears it on the belly and upper body of a sick man. The shaman is in a trance, chanting and possessed by his spirit friend. The patient recovered. (T. Headland, 1976.)

they often bury people within a few meters of where previous Agta have been buried. When it is time for the burial, the corpse is wrapped in a blanket, which in turn is wrapped in a woven grass mat and placed in a coffin. Coffins are made of split bamboo woven together in triangular fashion with the two sides tied together at the top and a bottom on which the corpse is laid (see Figure 8.2). As a step in the funeral ritual, everyone present gets in line and steps over the coffin and back again. It is then carried to the place of burial.

Bodies are buried in a supine position. As the body is being lowered into the grave, the surviving spouse may wail and roll on the ground. Sometimes he or she will try to prevent the coffin from being placed in the grave or will jump into the grave and throw out the dirt that is being placed over the coffin, but relatives will restrain the spouse. Important possessions may be buried with the body: bead necklaces, bow and arrows, cup and plate, clothes, and a cloth blanket. For infants, a coconut bowl with some of the mother's breast milk is left at the head of the grave. These items are to be used by the deceased in his or her next existence. A bamboo cross is placed at the head of the grave with bright-colored strips of cloth tied to it. A lean-to is built over the grave as protection from rain. Decorative flowers or plants may be planted around the grave. Then all return to camp.

As people enter the camp, a few women who had remained there will throw dust on everyone as they enter. This announces the end of the funeral. If the deceased is an adult, the house where the death occurred is burnt to the ground that evening. The next day the whole camp moves

Figure 8.2. Agta corpse in a bamboo coffin being carried to a burial site. (T. Headland, 1967.)

away from the rivershed for several weeks, and all of the other houses are abandoned or burnt. This is done out of fear of the spirit of the deceased, who may want others to accompany him or her in the afterlife.

LEVELS OF AGTA MORTALITY

Table 8.1 presents the profile of San Ildefonso Agta mortality employing the life table functions of q (probability of dying within a certain age range), l (probability of surviving to the beginning of an age range), e (life expectancy beyond a certain age range), and crude rates. The life expectancy at birth (e_0) for the research period (1950–1994) is 25 years, a very low life expectancy compared with those of industrial and some agricultural societies. Another way of expressing the same thing is the low survivorship rates (l). When a group of infants is born in a year, 29 percent die in their first year of life with only 71 percent surviving. By the time they are 10 years old, only half of them are still alive. By the age of 50, only a quarter are alive. This high level of mortality is also reflected in the crude death rate of 42.7 for the study population. The q rates are very high for all the age classifications, especially the early years of life. Two to three out of every 10 infants die in the first year of life.

REASONS FOR THE HIGH MORTALITY LEVELS

Table 8.2 lists the causes of the enumerated deaths. Causes of death can be difficult to determine in industrial societies, but in populations such as the Agta, the problems are compounded. Seventy-two percent of the deaths

Table 8.1. Agta Mortality Rates

Level of Analysis	Rate	Age	Total	Phase*		
				F	T	P
Individual	e_0		24.9	24.3	29.2	22.2
Age groups	Probability dying					
	q	0	292	369	241	266
		1	241	128	227	342
		5	69	48	55	99
		10	55	56	62	47
		20	85	97	78	81
		30	251	344	187	231
		40	251	290	155	311
		50	418	533	246	443
		60	697	506	852	641
		70+	1,000	1,000	1,000	1,000
	% survivorship to age					
	l	0	100	100	100	100
		1	71	63	76	73
		5	54	55	59	48
		10	50	49	55	44
		30	43	45	48	38
		50	24	21	33	20
Study population	Crude death		42.7	45.6	34.8	47.6
Number of deaths	Total		364	117	100	147
	Male		163	36	54	73
	Female		161	47	40	74
	Unknown		40	34	6	—

* F = Forager Phase; T = Transition Phase; P = Peasant Phase
q = probability of dying within a certain age range
l = probability of surviving to the beginning of an age range

were due to unknown causes, usually reflecting people's answers that the deceased took sick and died. This greatly restricts the analysis and prevents the drawing of firm conclusions.

However, several sources of information can be combined to formulate some tentative hypotheses about the factors behind the high levels of mortality. An examination of the 28 percent of known causes of death in Table 8.2 gives some indication of the main types of pathology. Further insight into the data of Table 8.2 can be gained from the incidence of morbidity indicated in Table 8.3, which reconstructs the frequency with which the Headlands treated various illnesses. More clues are provided by ethnographic observations of sanitation. Finally Table 8.1 will be reexamined. When certain types of pathologies are prevalent, they can be detected because they yield distinctive relationships between specific age segments of mortality.

CAUSE OF DEATH CATEGORIES IN TABLE 8.2

Table 8.2 gives the absolute and percentage values for five generic categories of causes of death. Although there is a high percentage of unknown causes, this does not necessarily imply that all the known causes are badly underenumerated. The enumerations of maternal deaths, accidents, and homicides should be good approximations. These are events in which the cause is easy to identify and of interest to the community, so that they are talked about and remembered. This is especially true of homicides, where the enumeration may be perfect. Therefore, based on the incidence of the two remaining known causes, it may be assumed that most of the unknown cases are infectious disease. Using this assumption to revise the percentage distribution of the causes of death, a revised column, "Assumpt.," was added to Table 8.2. For comparative purposes, an additional column ex-

Table 8.2. Causes of Death among the Agta, 1950–93

| Cause | Phases* | | | Total | | Assumpt. | Per Pop. | Per 10,000 |
	F	T	P	#	%	%	100,000	Live Births
Known	21	27	55	103	28.3			
Unknown	96	73	92	261	71.7			
Total	117	100	147	364	100			
Known								
Infectious disease	7	12	32	51	14.0	85.7	598	
Nutritional disease	2	3	10	15	4.1	4.1	35	
Maternal	7	4	5	16	4.4	4.4	188	352
Accident	3	7	—	10	2.7	2.7	117	
Homicide	2	1	8	11	3.0	3.0	129	
Total	21	27	55	103	28.2	100		

* F = Forager Phase; T = Transition Phase; P = Peasant Phase
Assumpt. = probable or assumed to be the causes of death, by percentage (see chapter 8 for explanation)

presses the magnitude of each cause as a rate per 100,000 population, a frequent demographic convention for expressing mortality rates. The last column shows the maternal mortality rate by yet another convention frequently used for it.

Infectious Diseases

Included under infectious diseases are measles, diarrhea, pneumonia, malaria, and tuberculosis. The infectious diseases comprise 50 percent of the known causes and 86 percent of all cases if the previously discussed assumption is accepted. Table 8.2 shows that the search for an explanation of the high mortality centers on infectious diseases. This is a broad category and frequently associated with parasitic and/or nutritional pathologies. Seldom are deaths attributed directly to parasites or malnutrition, even though it is known that these two factors are present and linked to infections. They operate concomitantly, and in many cases it is difficult to determine which is the primary cause, if indeed the concept of a primary cause has much relevance where these pathologies are operative.

Nutritional Diseases

This category was restricted to two types of cases, although it is known that nutritional problems may have been involved in many more cases of death. The first was beriberi, and there were three incidents of it, two in the Peasant Phase. Beriberi is caused by a lack of vitamin B_1. In the Forager Phase, the Agta received only unhusked rice in trade. They removed the outer husks by hand-pounding (see Figure 1.8), thus retaining most of the thiamine (vitamin B_1). By the end of the Transition Phase, the rice they bought or acquired in exchanges was milled (polished) rice. The milling process takes away much of the thiamine. Also, meat and fish were much more frequently eaten at meals in the Forager Phase than in the Peasant Phase. In a 1983 study the Agta had no meat or fish protein at 41 percent of their meals, whereas traditionally such protein would have been present at most of their meals (Headland 1986:457, 555). While these data are sketchy, they support the observation of a nutritional stress among the Agta in the Peasant Phase.

The other type of mortality included under nutrition is deaths attributed primarily to alcohol. Agta consume large quantities of alcoholic beverages. The following examples show how it results in death. A man who had been drinking heavily started back to his camp by following the shoreline. He passed out on the beach and remained there a whole day in the hot sun, becoming severely dehydrated before he went into a coma. When he was found, the family attempted to force liquids into him but without success, and he died six days later. In another case a woman stopped eating and spent a 24-hour period in heavy drinking until she passed out and never regained consciousness. In other cases, two tubercular Agta and two with pneumonia became drunk. All four went into comas and died.

Maternal Deaths

From 1950–1993, there were 16 cases where women died in childbirth. Many of these deaths appear to have been related to the poor health of the mother. This rate is more accurately and frequently expressed as per 10,000 live births. The Agta level of 352 is the highest we have been able to find in the literature. Other high levels are 42 for currently developing countries (Williams et al. 1994:179), 58 for Belgium 1903–10 (Meigs 1917:56), and around 150 for U.S. nonwhite population during the influenza pandemic of 1918 (Shapiro et al. 1968:145). In Headland's earlier (1989) research of the Casiguran Agta mortality, he found that 12 percent of the adult female deaths were from childbirth complications.

Homicide

There have been 11 homicides among the San Ildefonso Agta since 1950, 10 men and 1 woman. Eight of the deaths were at the hands of other Agta, two were by lowlanders, and in one case the assailant was unknown. In the eight cases where the assailants were Agta, the parties knew each other, and in four of them, they were kin-related. Five of the cases involved heavy drinking by the assailant and/or the victim. In one case two brothers-in-law were drunk and arguing about who was the better swimmer; one then shot an arrow into the arm of his in-law, who retaliated by stabbing his assailant five times with his bolo. In another case a woman was beaten to death by her husband, who then threw her body off a cliff into the ocean. The Agta have a long history of violence among themselves, including intergroup raiding (Headland 1986:385–88, 391–93).

The in-migration of lowlanders has increased the number of lowlander-Agta disputes. In 1990 a young Agta found lowlanders cutting a valuable hardwood tree on his father's land and protested. The lowlanders attacked him, and he retreated. However, he returned two hours later, shot and wounded one of the lowlanders in the back with a barbed arrow, and then fled. A month later, the young Agta was murdered in his sleep by an unknown lowlander assailant. One blow with a bolo had sliced open his upper chest cavity. (Six days later the victim's relatives made a revenge raid against a lowlander homestead at Dinipan; they ambushed a man there, filled his body with nine arrows, and fled.)

The San Ildefonso Agta homicide rate of 129 appears moderate for anthropological populations. It is well below the rates for certain New Guinea and Australian groups and slightly below the rate for the Amazonian Yanomami, and less than half the rate for the Casiguran Agta on the mainland. (See the tables in Knauft 1987:464; Headland 1989:69).

Accidents

There were 10 fatal accidents during the study period. In one of three drowning incidents, a family was paddling across the bay in a dugout canoe. The man, who was intoxicated, had a fishing net with lead weights draped over his shoulders. He stood up in the boat to urinate. The boat tipped over, and with the weight of the net holding him down, he never surfaced. In another case a young child was accidentally hit with a stick by another child and died a few days later. Another case was a hunting accident. Other accidental deaths include a man hit by a falling tree and another who fell out of a tree. One ate a poisonous crab and two were somehow related to fishing accidents. All were males. There is no record of an Agta ever dying from a poisonous snake bite or by a python's constrictions.

MORBIDITY

Another indication of the causes of the high death rates comes from the types of symptoms and sicknesses the Headlands observed in their paramedical program. The Headlands estimated how often they treated specific symptoms and illnesses. Table 8.3 tabulates the results by stating how many cases were treated per week or per multiple weeks. While the incidence of morbidity is not necessarily correlated with the incidence of mortality, the table does indicate certain types of pathology. Parasites were the most frequently treated complaint. Coughs, fevers, colds, pneumonia, bronchitis, and tuberculosis indicate respiratory infection. Diarrheas show gastrointestinal infection. Nutritional stress in the population is also indicated. The poor health and nutritional conditions indicated by this table are probably responsible for the late age of menarche for Agta girls mentioned in chapter 7.

CAUSES OF MORTALITY IN THE PEASANT PHASE

Some Indications

The known causes of death have shown the importance of infectious diseases in Agta mortality. The role of these diseases has been confirmed by their importance in the types of morbidity treated by the Headlands. Mal-

Table 8.3. Symptoms, Treatment, and Estimated Case Load: Headlands' Paramedical Program, 1962–79, 1983–86

Symptoms	Treatment	# Cases	Every # of Weeks
I. Minor Complaints			
Intestinal parasites	Piperazine, Antiminth P.	6	1
Coughs	Cough syrup	4	1
Fever	Aspirin	4	1
Colds	Aspirin	2	1
Cuts, open sores	Hydrogen peroxide	2	1
Minor diarrhea	Kaopectate, Lomotil	2	1
Stomachache	Paregoric	1	1
Conjunctivitis	Antibiotic cream	1	1
Rheumatoid arthritis	Daily aspirin	1	1
Infected sores	Hydrogen peroxide, antibiotics	3	2
Scabies	Medicine, wash clothing, etc.	1	3
Toothache	Aspirin	1	5
Boils	Compresses and aspirin	1	17
II. Critical, Nonemergency			
Malnutrition	Multivitamins, food portions	2	1
Malaria	Chloroquine	1	2
High fever	Aspirin, antibiotics	1	2
Severe diarrhea	Kaopectate, antibiotics, rehydration	1	3
Tuberculosis	INH, Streptomycin	1	3
Anemia	Iron, multivitamins	1	5
Bronchitis	Cough syrup, antibiotics	1	10
Major stomachache	Paregoric	1	25
Asthma	Antiasthmatics	1	100
Leprosy	Diaminodiphenylsulfone	1	520
III. Critical—Emergency			
Pneumonia	Antibiotics	1	10
Severe malnutrition	Food supplements, multivitamins	1	13
Severe stomachache	To town doctor	1	20
Severe dehydration	Rehydration	1	25
TB with blood vomiting	Vitamin K	1	33
Diseased liver	To hospital	1	52
Convulsions, fever	Diazipam, antipyretics	1	52
Snake bites	Antivenin injections	1	156

The next-to-the-last column indicates the number of times a treatment was done during the number of weeks shown in the far right column. For example, the number of times treatment was administered for intestinal parasites averaged six times per week.

nutrition has also been shown to be important by the morbidity seen in the treatment program as well as the poor economic conditions of the Agta in the Peasant Phase. Another possible indicator is the drop in the consumption of meat and fish shown by the Headlands' dietary survey. Infectious disease and malnutrition can enter into a synergism so that the pathology they produce has a multiplier effect rather than being additive, that is, the sum of each as if working independently.

The Synergism of Malnutrition and Infectious Disease

The synergism takes time to take its toll. An infant will suffer an infectious disease, frequently diarrhea. The infant will usually recover from the first such attack, but if malnutrition is present, this is not a full recovery to good health. The infant remains in an undernourished state, and its general health continues to deteriorate, which leaves the child susceptible to another attack of infectious disease. Again, it may recover from the infectious disease, but its health condition has been weakened even more, and the malnutrition continues. Finally another attack of any kind of infectious disease may simply overwhelm the already weakened infant, with the resultant mortality. As Salomon, Mata, and Gordon (1968) have noted:

> The effect of the communicable diseases of childhood—measles, whooping cough, mumps, rubella, and chicken pox—on nutritional status of the patient scarcely rests in the episode itself, but rather in the extent the attack is part of a sequence of repeated, often nonspecific infections, most often the ordinary infectious syndromes of intestinal and respiratory tracts. The effect is additive and cumulative. Conversely, these usually benign diseases derive their enhanced fatality, greater frequency of complications and exaggerated clinical course through attack on a less resistant host, to which nutritional deficiency contributes importantly. To know the full potentiality of these diseases in developing countries, information is necessary on preceding numbers and duration of these other infections.

This pathology is predominant among peasantry in India and Guatemala, where Scrimshaw, Taylor, and Gordon (1968) did their seminal work on the synergism. In Guatemala it has also been studied clinically by the Instituto de Nutrición de Centro América y Panamá (INCAP) and demographically by Gordon et al. (1967) and Early (1982). This work has yielded several distinctive demographic indices that can be used to detect the presence of the synergism. The question is: Was the infectious disease and malnutrition indicated in Tables 8.2 and 8.3 of such severity that they triggered the synergism? If so, the synergism was an important factor in the high level of mortality in the Peasant Phase, as shown in Table 8.1.

Demographic Indicators of the Synergism

A common demographic assumption is that the human mortality rate is constantly falling through the first year of life and continues to decline over the next four years. This is the profile in industrial populations, where the mortality rate drops sharply after the first week of life and continues to decline for the remainder of the interval. Within the first year of life, neonatal mortality rates (first month or 28 days of life) primarily reflect pathology connected with pregnancy or the birth process. Postneonatal rates (remainder of the first year of life) primarily reflect the pathology of the environment in which the infant exists after birth. Gordon, Wyon, and Ascoli (1967) noted that the magnitude of the decline between these two segments of the first year death rate (the infant rate) in developed countries is expressed by a 2 to 1 ratio.

However, Gordon, Wyon, Ascoli (1967:360, 366–67), and Early (1982:118–23) have shown that in peasant areas where the synergism is operative a different and distinctive age pattern of mortality emerges in the first five years of life. These distinctive patterns will be examined and then used to test the presence of the synergism among populations lacking clinical studies, such as the Agta.

Within the First Year of Life

The distinctive pattern first shows up in the internal structure of the infant mortality rate. The pathology begins to build during the first year of life, so that by the twelfth month, Gordon et al. have shown that the postneonatal death rate is greater than the neonatal rate by nearly a 2 to 3 ratio or more. This is the reverse of the expectation for industrial populations. Therefore ratios of this structure and magnitude become demographic indicators of the presence of the synergism.

A more sensitive indicator for the first year was developed by Early. He took Gordon's work on the infant mortality rate a step further by noting that the mother's milk usually provides sufficient nourishment for the first six months of life. If sufficient and proper supplements are not given after this, the synergism begins to build, with the resulting mortality. Therefore the infant death rate can be divided into three segments: the neonatal (the first month), the second to sixth month, and the seventh to the twelfth month. Where the synergism is dominant, a graph of the magnitudes of three segments is V shaped, that is, the one-to-six-month segment is less than both the neonatal and the seven-to-twelve-month segments. The mother's milk holds down the two-to-six-month segment.

The Second Year of Life

Proceeding into the second year of life (the 1 year), the human mortality curve in industrial populations continues its decline. In populations where the synergism is active, the mortality curve continues to increase its magnitude. The value of the second-year death rate reaches at least 50 and can climb to over 200 per 1,000 population. Values of these magnitudes mean that the second-year death rate may be greater than the postneonatal rate of the first year, and this relationship becomes another demographic indicator of the synergism.

Decline of the Mortality Rate

At what age does the mortality curve peak and begin its descent in populations marked by the synergism? Epidemic periods need to be distinguished from nonepidemic years, that is, years of endemic malnutrition without epidemics of infectious diseases. In nonepidemic years, there are indications that the curve peaks in the 12–17-month period of the second year of life (the 1 year) and then begins its decline. In epidemic years, the decline does not begin until the third or fourth year of life (the 2 or 3 year), depending on the type of infectious disease. In contrast, the mortality curve of industrial systems continues the decline that began after the first week of life.

Combining the 1 to 4 Years

Because of the high mortality in the second year of life and in epidemic years continuing into the third and fourth years, the mortality rate for the combined ages of one to less than five years is high in populations suffering from the synergism. It is so high that the value of the mortality rate from one to less than five years ($_4q_1$) is greater than the infant mortality rate for the same period. This indicator of the synergism is useful because of the age classifications frequently used to publish demographic data.

Which Indicator?

Which is the best demographic indicator of the synergism? Gordon et al. (1967:367) opt for the magnitude of the second-year death rate because they find it is less influenced by other factors than any of the other indicators. A strong argument can also be made for the presence of the synergism when all the previously mentioned indicators meet the criteria for its influence.

The Agta Indices

The first part of Table 8.4 shows the demographic values for Agta mortality in the first five years of life during the Peasant Phase. Of the 169 births in this period, there were 45 infant deaths and 37 in the remaining four years. The table indicates that the Agta mortality rates have the structure and magnitude that meet all the criteria for the presence of the synergism.

However, there is a problem with the data used for the mortality profile in the first part of the table. The criteria require the use of age classifications by individual years and within the first year by months. To obtain accurate ages of death for such discriminate intervals requires good registration of the dates of both births and deaths. Tables 4.1 and 4.2 show that about 45 percent of the births and deaths in the Forager Phase were either estimated or exact only as to year but not month and day. Some self-cancelation is assumed to have taken place so that estimation error is mitigated to some extent. But this still leaves questions about the validity of the analysis in the first part of Table 8.4.

To gain greater control over the accuracy of the ages at death, a similar population was defined and its mortality examined. This population, called

Table 8.4. Agta Mortality Indices per 1,000 Population for the First Five Years and the Criteria Indicative of the Synergism of Malnutrition and Infectious Disease

Mortality	Peasant Phase 1980–93			Criteria of Synergism	Age Sample		
	Births	Deaths	Rate		Births	Deaths	Rate
First Year	169	45			103	31	
$_1q_0$			266				301
Neonatal		20	118			10	97
Postneonatal		25	148			21	204
Ratio		2:2.5		nearly 2:3		1:2.1	
Months:							
1 : 2–6 : 7–12		20>10<15		1>2–6<7–12		10>9<12	
Second Year		19					
$_2q_1$			153	50+			
Relation to postneonatal			153>148	$_2q_1$>postneonatal			
Second to Fifth Years							
$_4q_1$		37	342				
Relation to infant rate			342>266	$_4q_1>_1q_0$			

the Age Sample in the second part of Table 8.4, consisted of any birth in the Agta database between 1979 and 1994 in which the date of birth was accurate at least as to both year and month. The mortality of this population with 43 percent dead proved to be similar to that for the Peasant Phase with 49 percent dead. However among the deaths of the age sample, 88 percent of them had dates of death accurate at least as to year and month while only 54 percent of the deaths in the Forager Phase had this degree of accuracy. With this greater control over the age at death, the age sample shows all the criteria for the presence of the synergism are verified in the first year of life. This shows that estimation error of ages is not invalidating the mortality profile in the first part of the table. The number of cases in the age sample became too small for meaningful analysis of the other ages.

Synergism Summary

This section has described previous demographic work by Gordon, Wyon, Ascoli, and Early among peasantry whose mortality was heavily influenced by the synergism of malnutrition and infectious disease. This work has established the distinctive demographic magnitudes and relationships characteristic of the mortality profiles generated by this type of pathology. These criteria were applied to the Agta mortality data for the Forager Phase. The Agta data met all five criteria. However there could be a question about the validity of this finding because of the exactness required for the ages at death. Another population, the age sample, was defined, and it contained a high degree of accuracy for the ages at death. In the first year of life, all the criteria for the presence of the synergism were also met.

Therefore, these distinctive magnitudes and internal relationships of the Agta mortality rates for the first five years of life indicate that the Agta are suffering from the synergism in the Peasant Phase. This is due to the loss of their foraging habitat and being forced into the rural labor market, where they obtain intermittent work and extremely low wages, which in turn result in dietary deficiency. These conditions are compounded by the infectious diseases brought by the lowlander in-migrants and the much denser population in which the infectious diseases have more opportunity to find hosts (Black 1975).

Although the Agta data confirm the presence of the synergism, in some cases the criteria were just barely met. The data do not have the overwhelming levels of the data from India and Guatemala. This difference may indicate a lesser impact of the synergism on the Agta than among these other peasant populations.

Causes of Mortality in the Forager Phase

So far the analysis has shown that parasites, infectious disease, and malnutrition are the main causes of Agta mortality in all three phases. Table 8.1 shows that the life expectancy at birth and the crude death rates for the Forager and Peasant Phases are almost identical. The demographic indicators have shown that in the Peasant Phase, the synergism of malnutrition and infectious disease was present and thus an important factor in the level of mortality. Does this mean that the synergism of infectious disease and malnutrition was also operative during the Forager Phase?

In Table 8.1 the infant mortality rate of the Forager Phase is greater than the child mortality rate (ages one through four), a difference that would appear to be a demographic indication of the absence of the synergism. But there is a methodological problem with these rates and any conclusion drawn from their relationship. In the construction of the fertility histories, it was known that a number of births had taken place during the 1950s and the infants had died very young, presumably in the first year of life. But some of them may have died in the second or third year of life. Therefore estimation error for ages appears to be part of the reason for the very large differential between the probability of dying in the first year of life (q_0) and the probability of dying between the ages of one to four $({}_4q_1)$ in this phase. The other demographic tests used for the Peasant Phase cannot be used for the Forager Phase for the same reason—lack of exact dates of birth and death for the same individuals (see Tables 4.1 and 4.2).

The ethnographic evidence indicates dietary sufficiency in the Forager Phase. The San Ildefonso Agta had sufficient game, fish, and collected foods for consumption and to exchange for rice with the lowlanders. Peterson (1978:83), Rai (1990:115), and Griffin et al. (1992:176) found the same thing among the Agta of Isabela and Cagayan. There were times of shortages such as those caused by damage from typhoons, but these were not constant conditions. During the Forager Phase there were no known epidemics of infectious disease nor was there endemic malaria. Agta are cognizant of epidemics and know the Tagalog word for it. In the 1970s the Headlands interviewed many Agta about this subject; several remembered an epidemic disease among wild pigs, but none could recall any epidemic disease causing deaths among Agta. Also the population of the San Ildefonso Peninsula was much smaller in this phase, and this lower density offered fewer opportunities for some infectious diseases to survive. Therefore, although the overall mortality levels are almost identical, the ethnographic evidence suggests that the structure of the mortality may have been somewhat different in the Forager Phase than in the Peasant Phase.

The adult mortality in the Forager Phase is higher than in the Peasant Phase. The probability of dying (q) rates for this phase are the highest from ages 10 through 59 except for the 40s, only slightly lower than that of the Peasant Phase. Also the combined infant and childhood mortality rates of the Peasant Phase are slightly higher than those for the Forager Phase, as shown by the survival (l) rate to age five. Fifty-four percent of the live births are still alive at age five in the Forager Phase while only 48 percent are in the Peasant Phase.

A possible clue to the pathology of the Forager Phase may be the heavy parasitic loads found by the Headlands immediately upon their arrival in 1962 and their constant treatment of them (Table 8.3). Poor sanitary conditions among the Agta favor the spread of parasites. Adults and older children defecate 30 to 100 meters from the camp. Hands are not washed after defecating, and food is eaten with the fingers. Tin plates and cooking utensils are placed unwashed in the roof thatch. Inside the small houses, the dirt floors and the sleeping mats kept on them are frequently soiled by infants and toddlers. In the warmer months, April through September, Agta bathe almost daily, but seldom do so in the colder season (October–February). These conditions are not unique to the Agta but are present in many societies that are unaware of the pathology generated by the micro world.

As a result, Agta camps have many insects, especially cockroaches, flies, and mosquitoes. Intensity varies with time of year, frequency of camp moves, and the degree of care of the house site by individual families. As mentioned in chapter 1, the Agta are not nomads. Two studies (Rai 1990:58 and Clark 1990:13) showed some Agta groups move their camps an average of every 18 or 37 days, which allows time for parasite buildup. These conditions are favorable for the ingestion of internal parasites—hookworms, whipworms, pinworms, and roundworms. At the time of a child's death, roundworms may exit from the bodily orifices, primarily the mouth and nose. These conditions also favor external parasites, especially mites that cause scabies, which can be extremely irritating, leading to excessive scratching and infected sores.

From these clues, one can speculate that mortality from parasitic pathology may have played a somewhat greater role in the Forager Phase than in the Peasant Phase, although parasitic loads were also heavy in the Peasant Phase. Morbidity from parasites may last for a longer time before it results in death. In the Peasant Phase parasitic disease may not get as much opportunity to take its toll because the human host is cut down earlier by the synergism of malnutrition and infectious disease, to which the parasites also make their contribution. There are probably synergisms created by heavy parasitic infestations. We need to know much more about the types of parasitic diseases and their developmental cycles.

Even with dietary sufficiency, malnutrition may have been a problem in the Forager Phase. Parasites themselves divert nutrition from the host. Also the Agta's excessive use of alcohol contributed to nutritional problems in all three historical phases. It creates nutritional problems not only for the drinker but also for those dependent on his or her cooperation in subsistence activity. Some may speculate that the Agta's use of alcohol is a reaction to their losing their traditional habitat and way of life. However, the excessive use of alcohol appears to be an Agta custom going back to the group's foraging days. Turnbull (1929:209; 1930:92, 783) noted it in the second decade of the century, and Vanoverbergh (1937–38:924) mentioned it as a serious problem in 1936.

The Decline of Mortality in the Transition Phase

Table 8.1 shows that in the Transition Phase there was a decrease in mortality from the Forager Phase—a five-year jump in the life expectancy at birth and a 24 percent drop in the crude death rate. What factors were responsible for this decline?

Table 8.3 indicates the medical program of the Headlands and also medical assistance rendered by several teams of the New Tribes Mission as important factors (see Figure 8.3). The efforts of the missionaries definitely saved some lives, but their exact impact cannot be measured. The Headlands were living among the Agta for extended periods, spoke the language, and were trusted. Not only had they received some paramedical training before going to the Philippines, but they were in daily contact by radio with a physician stationed near the SIL center in Bagabag, Nueva Vizcaya, as well. They began treating Agta soon after their arrival in 1962 and expanded treatment during the late 1960s and 1970s, but had to neglect it from 1979 to 1982 owing to their absence from the Philippines. They terminated the program with their departure in 1986. Therefore, its main effect was during the Transition Phase.

Another factor during the Transition Phase was the increase in health services available in the town of Casiguran. Two Ilokano physicians, a husband-wife team, began their medical practices in the town of Casiguran in the 1960s. While they did not reach out to the Agta, the Headlands and the New Tribes missionaries brought some of their critical patients to those doctors. The delivery of health services took a big step forward with the opening of a government hospital in the town in 1977. Formerly the only hospital services available were in Baler more than 80 kilometers (52 miles)

Figure 8.3. Thomas Headland injects vitamin B complex into an Agta woman suffering from malnutrition. She is in her house. (J. Headland, 1971.)

distant by trail from the Casiguran area. The Headlands believe that one of the main contributions of their program was creating trust among the Agta toward these health services. By 1986, when the Headlands left the Philippines, most of the Agta were aware of—and usually no longer afraid to go to—the hospital to use these services on their own initiative.

THE RISE OF MORTALITY IN THE PEASANT PHASE

The decline of mortality during the Transition Phase was reversed during the Peasant Phase and returned to the previous level of the Forager Phase. Life expectancy at birth dropped to 22 years in the Peasant Phase, and the crude death rate rose to 47.6. The Agta no longer had medical assistance from the Headlands. Two women of the New Tribes Mission lived with Agta on the peninsula from 1980 to 1992 and provided some medical help, but they were unable to overcome the increased sources of disease the Agta now faced. The Agta by this time depended to a great extent on wage labor, which was intermittent. Their diet suffered, and the synergism began to build. This Peasant Phase was marked by the huge increase in the lowlander population and the immersion of the previously semi-isolated Agta into the middle of these in-migrants. The Agta were subject to the infectious diseases carried by the newly arrived lowlanders. The higher population density provided more hosts with low immunity. There had never been an epidemic among the Agta in this century until the Peasant Phase. In 1985 there was an outbreak of a strong diarrhetic disease, resulting in 16 deaths; the crude death rate soared to 71. In 1987 there was a measles epidemic. During March and April of that year, 23 Agta children died from measles, and the crude death rate reached 95. Mosquitoes had always been present on the peninsula, but malaria had been rarely seen until this phase.

Another factor contributing to the synergism was the adoption of bottle-feeding by some Agta mothers. Breast milk enables an infant to ward off infection by receiving antibodies from the mother. In a 1983 survey the Headlands found 10 suckling children in the Casiguran area being bottle-fed, approximately 20 percent of all Agta sucklings. In a follow-up study two years later, only three of those bottle-fed children were still alive. Two main reasons for bottle-feeding were given by the mothers: Some said they did not have enough breast milk; others said that lowlanders had told them canned milk was better for their infants. Prestige is probably another factor. A number of lowlander mothers bottle-feed their babies, and the Agta were emulating them. But the practice is lethal for the Agta. They do not properly wash, much less sterilize, the bottles and nipples, nor can they refrigerate the milk. And quite often the so-called milk sold to the Agta is not really milk but rather a heavily sugared liquid made from coconut or vegetable oil mixed with a small amount of nonfat skim milk. It is condensed and needs water added, but the water is seldom boiled and frequently too much is added in an effort to make it last longer. This misuse compounds the infant and child mortality.

AGTA HEIGHT AND WEIGHT

Table 8.5 presents Agta weights in kilograms, heights in centimeters, and the ratio of the two multiplied by 100. The heights have remained much the same over the 18-year span of the table, while there has been a slight gain in the weights. The total ratio for all time periods for males is 29.9 and for females 27.3. These figures are comparable to other studies of Asian Negritos, African Bushmen, and Pygmy groups, but there is disagreement about their significance. On the one hand, Tobias (1964:76) and Lee (1979a:289–92) see these peoples' small build as efficient adaptation to a foraging life in a particular climate. On the other hand, Truswell and Hansen (1976), Griffin (1984b:22–23), Headland (1986:394–96), and Eder (1987:143) find problematic nutrition as the responsible factor for the group members' small stature and thin physique. There is no agreement on what standard should be used when it is said these groups fall below the "standard." Pennettii et al. (1986:129) grant the nutritional argument but see a

Table 8.5. Agta Weights (kg) and Heights (cm), 1976, 1983, 1992, 1994

	Total	N	1976	1983	1992	1994
I. Weight (kg)						
Total	42.7	366	41.1	42.1	42.8	43.7
Male	46.2	178	45.6	45.5	46.3	47.0
Female	39.3	188	37.4	38.6	39.8	40.5
II. Height (cm)						
Total	148.8	236	149.9	147.8	—	149.1
Male	154.4	116	155.4	152.7	—	154.9
Female	143.8	120	144.3	143.3	—	143.8
III. Ratio—Height/Weight						
Total	28.7		27.4	28.5	—	29.3
Male	29.9		29.3	29.8	—	30.3
Female	27.3		25.9	26.9	—	28.2

possibility of additional genetic factors. This research is only able to add some speculation about the factors that may have been important in the Forager Phase. The weights and heights are presented here for comparative purposes but have not been used in the analysis.

SUMMARY

The Agta are a high-mortality population with a life expectancy of about 25 years. The main problems appear to be parasitic loads, infectious disease, and malnutrition. For the Forager Phase, data are sparse on the causes of the high level of mortality, and only speculative conclusions can be drawn. Parasitic infections as a result of unsanitary conditions may be an important factor. In the Peasant Phase, the relationships between segments of the mortality rates for the first five years of life indicate the presence of the synergism of malnutrition and infectious disease as an important factor. This is reinforced by the demographic information about the size of low-lander in-migration, which would be an additional source of infectious disease. Malnutrition appears influenced by the decline in fish and wild game as well as the low income, because of the place of the Agta in the rural Philippine labor market.

Chapter 9

Migration

In the Forager Phase, the Agta were accustomed to moving their camps long or short distances for work, family, or other reasons. Their camps were made of simple structures that could be easily built in a day and just as quickly abandoned. Even today in the Peasant Phase, when the Agta seldom forage and have become semisedentary, they retain the custom of easily moving for whatever reason that may arise.

As explained in chapter 4, not all moves are considered migrations. For a move to be considered a migration for a family group, they must remain on or away from the peninsula for at least three years. For an individual moving without a family group, a one-year presence or absence is required. Defined in this manner, from 1950 to 1993 there have been 485 in-migrations to the peninsula and 523 out-migrations, for a net migration of -38.

Table 9.1 presents the crude rates of in-migration, out-migration, and net migration along with their absolute values. These are high rates of 50 or more, comparable to the levels of the crude birth rates. These levels of migratory rates are unusual but probably characteristic of most foraging populations defined by a geographical area. Despite the high levels, in-migration and out-migration approximate each other so that net migration remains relatively small.

LEVELS OF MIGRATION AND DEFINITIONS OF POPULATIONS

The levels of the rates of migration are partly dependent on the way the population is defined. In studies using residence as a criterion, these defini-

Table 9.1. Crude Rates and Number of Migrations, 1950–93, by Phases

Rate	Total	Phase*		
		F	T	P
In-migration	56.9	56.5	48.7	64.8
Out-migration	61.3	52.2	72.4	58.6
Net migration	-4.5	4.3	-23.7	6.1
N				
In-migration	485	145	140	200
Out-migration	523	134	208	181
Net migration	-38	11	-68	19

* F = Forager Phase; T = Transition Phase; P = Peasant Phase

tions set the outer limits that need to be crossed for a migration to take place. Usually the larger the area of the defined population, the lower the level of migration and vice versa. The larger area means that more of the movement is internal to the population, not migratory in and out of it. The levels of migration can be quickly altered by differing definitions of the area limits of a population. This study has been restricted to the Agta of the San Ildefonso Peninsula for the reasons discussed in chapter 4. If a study of Agta over a larger geographical area had been possible, the levels of in-migration and out-migration would probably have been lower.

REASONS FOR MIGRATION

Table 9.2 examines the reasons for migration. The analysis is hindered by a lack of information about a substantial proportion of the Agta; 54 percent of the in-migrations and 31 percent of the out-migrations. The difference between the two percentages reflects the fact that it was easier to know the situation of families or individuals on the peninsula before they left than to ascertain the backgrounds of in-migrants. For some of the latter cases, "Return Home" has been used as a category. It was only known that an Agta family or individual was returning to the river valley where they more frequently resided, not the original reason for the move. In spite of these shortcomings, some insight into migration can be obtained from the known cases.

Marriage

The enumeration of marriage histories is almost complete as marriages are public knowledge and widely discussed in the community. A description of

Table 9.2. Number and Percentage of Reasons for In-migration and Out-migration

In-migration Reason	Total	Number*		
		F	T	P
Marriage	58	19	19	20
Family	32	11	6	15
Economic	11	2	4	5
Conflict	2	1	1	—
Return home	119	21	48	50
Unknown	263	91	62	110
Total	485	145	140	200

In-migration Reason	Total	Number*		
		F	T	P
Marriage	12	12	14	10
Family	6	8	4	8
Economic	2	1	3	3
Conflict	1	1	1	—
Return home	25	15	34	25
Unknown	54	63	44	55
Total	100	100	100	100

Out-migration Reason	Total	Number*		
		F	T	P
Marriage	67	24	21	22
Family	56	21	11	24
Economic	22	10	—	12
Conflict	20	1	11	8
Reservation	71	1	70	—
Maid	34	3	13	18
Return home	92	32	24	36
Unknown	161	42	58	61
Total	523	134	208	181

Out-migration Reason	Total	Number*		
		F	T	P
Marriage	13	18	10	12
Family	11	16	5	13
Economic	4	7	—	7
Conflict	4	1	5	4
Reservation	14	1	34	—
Maid	7	2	6	10
Return home	18	24	12	20
Unknown	31	31	28	34
Total	100	100	100	100

* F = Forager Phase; T = Transition Phase; P = Peasant Phase

the way marriages are arranged was given in chapter 7. One of the negative precepts of Agta marriage is that no one can marry a person whom that Agta already calls by a kin term, either consanguineous or affinal. A camp group is composed of related people, so usually everyone in one's camp group is eliminated as a potential marriage partner. Camp groups in proximity to each other are often related. The Agta population of the peninsula during the study period was small, averaging about 200 people, so there may have been difficulty finding someone of marriageable age who was not called by a kin term and who was not already married. Consequently the marriage rule stimulated migration. The San Ildefonso Agta have had marriage ties for many years with other Agta communities, especially on the Casiguran mainland, in Palanan, 90 kilometers (56 miles) to the north and in Madella, 40 kilometers (25 miles) to the west (Map 2). (The elderly man in the introductory photograph spent most of his life in Palanan and later in-migrated to live with some of his family who were married to San Ildefonso Agta. During World War II he and his family in Palanan hid a downed American pilot named Ronald Gower from the Japanese.)

There are two kinds of in-migration and out-migration for marriage reasons: physical migration, which has just been discussed, and cultural migration. Cultural migration takes place when an Agta marries a lowlander or a member of the Acculturating population (to be discussed in chapter 11) and the couple lives in a lowlander or Acculturating household. Frequently this marriage does not involve physical movement away from the peninsula but does mean a move from an Agta camp group to a lowlander hamlet on the peninsula. Table 9.3 shows the distribution of migration for marital reasons. Of the 45 female out-migrations for marriage, almost 50 percent were with lowlanders or other members of the Acculturating population. Although there were more Agta in-migrating for marriage to San Ildefonso Agta than San Ildefonso Agta leaving for marriage to outside Agta (58 to 39, for a net gain of 19), the loss of 27 San Ildefonso Agta by marriage to the Acculturating population resulted in an overall loss of eight people to the population.

The Calabgan Reservation

The government's Agta reservation in Calabgan (see Map 1) was started by U.S. Army Captain Turnbull in 1912 (see chapter 1). The reservation was never permanently occupied by the Agta but was periodically used by unsuccessful government programs for this purpose. It was the destination of the largest single Agta out-migration in the middle 1970s. These were the years when the forces of the New People's Army (NPA) brought guer-

Table 9.3. Migration for Marriage—by Sex and Population of Partner, 1950–94

	N	For Marriage To	
		Agta	Acc. Pop.
In-migration	58	58	—
Males	18	18	—
Females	40	40	—
Out-migration	66	39	27
Males	21	16	5
Females	45	23	22
Net migration	-8	+19	-27
Male	-3	+2	-5
Female	-5	+17	-22

rilla warfare to the Casiguran area, as described in chapter 2. Lowlanders congregated in the town of Casiguran to avoid the NPA. Many Agta followed them and camped on the outskirts of the town for safety. In 1975 the army moved most Agta to the old reservation site in Calabgan. Table 9.2 shows that 71 Agta moved from the peninsula to the reservation during this period. Only 46 of them, along with eight other in-migrants, returned immediately after the NPA guerrilla warfare. Others eventually migrated back to the peninsula after some years of residence elsewhere.

Maids

Another source of out-migration is Agta girls who leave the peninsula to work as housemaids for lowlanders. A total of 33 Agta girls did this during the 44-year research period. (One left twice, which made 34 out-migrations.) Thirty-eight percent of these girls went to the town of Casiguran and the rest to Manila or provincial towns elsewhere in Luzon. These maids live in lowlander houses where they work, learn Tagalog, and become accustomed to Filipino ways. The enumeration of this category is probably complete, as the absence of these girls is remembered in the Agta community. Most leave between the ages of 12 and 18 (the average is 15.3, and the median is 13) and stay away for an average of three years. Of the 33 out-migrants, 23 have returned to the peninsula: 19 immediately after working as a maid, and the other four after marrying and living in another community for a period of time. The immediate returnees averaged 18.6 years of age upon coming back. Ten have not returned to the peninsula; four of these are girls who recently left and probably will return; four are now older women who married while outside the peninsula and are living in other communities; and two are young women who were kidnapped while young girls in 1984 and are being held as bondage maids somewhere in Manila.

Family

This category reflects migration necessitated by a restructuring of family arrangements occasioned by divorce or death. The high mortality rate seen in the previous chapter means that there are numerous widows, widowers, and orphans who migrate to live with other family members outside the peninsula and vice versa. Frequently a later remarriage means another migration for a family group.

Conflict

This reason for migration refers to moves to resolve various types of disputes. The conflict may be between members of a camp or river valley group. Or it may be a dispute with an abusive lowlander who is trying to get an Agta family to work for him in repayment of a debt.

Economic

There are many economic reasons for Agta to migrate, as they recognize better opportunities elsewhere in hunting, wage labor, trading, and so forth. This reason for migration is probably highly underreported and includes many of the unknown cases. The most frequent reason for Agta migration is economic.

SUMMARY

This chapter has examined two of the demographic variables, in-migration and out-migration. The rates of both are extremely high, which would be expected for a foraging population and the small geographical area by which the San Ildefonso Agta population is defined. However, for most periods the two rates approximate each other so that the rate of net migration is relatively small. The largest known source of unbalanced migration movement was the guerrilla warfare in the middle 1970s. A significant number of Agta out-migrated then to the Calabgan reservation and did not return from there after the guerrilla warfare. They either stayed on the mainland or went elsewhere. Agta girls going to Casiguran, Manila, and other towns to work as maids were an important source of out-migration. Although they usually returned to the peninsula eventually, this migration became significant because these Agta learned to emulate Filipino culture.

Chapter 10

Some Characteristics of the Agta Population

The last three chapters have discussed fertility, mortality, and migration, the demographic components of population change. This chapter discusses the relationship of these factors to some characteristics of the Agta population. It first looks at the relationship between mortality and the Agta life cycle, and then between mortality and some Agta behavior patterns. Then the chapter examines the stability of the population structure in spite of significant ecological and social changes. Finally the chapter considers some factors that could undermine this stability.

THE AGTA LIFE CYCLE CONFRONTING THE HIGH MORTALITY

The Agta are a population of very high mortality, as shown by a life expectancy at birth of 25 years. The research in this book offers the opportunity to examine the demographic impact of the high mortality on the life cycle of the survivors and may be helpful as a basis for clinical and psychological studies. This type of high-mortality life cycle is distinctively different from the life cycle of survivors in low-mortality groups, where the occasions for the consciousness of death are delayed in the life cycle. Fewer of their children die, and close contemporaries die at an older age.

Facing the Mortality of the Self

The high mortality levels mean that from the earliest stages of the life cycle the realistic threat of one's own death is continually before the Agta consciousness. Figure 10.1 illustrates this by graphing the death and survival functions of the life table (see Table 8.1). It shows that an Agta has only a 50 percent chance of staying alive until age 10 and about a 25 percent chance of surviving to age 50.

But in this high-mortality population, it is not only a question of an early consciousness of one's own mortality. There is also the continual impact on this consciousness of the death of others. These include one's siblings and age mates, parents, spouses, and children. In terms of Figure 10.1, it is not only a question of when the self joins the dark-shaded space. There is also the stress of remaining in the light-shaded area while watching it shrink and the dark-shaded space expand. Like people the world over, Agta adults have a chronic fear of sickness and death (Headland 1987a:349). Some Agta fear this more than others do, of course, depending on the current health or danger of an individual or his immediate family. They talk about death more than do westerners, probably because they witness it much more often.

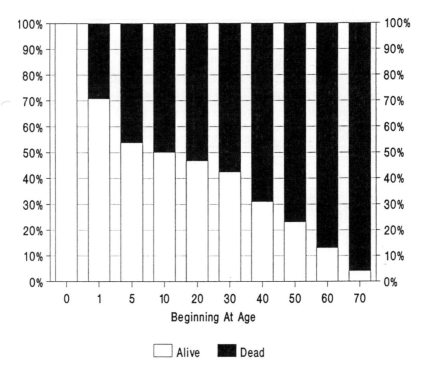

Figure 10.1. Percent dead or alive, by age.

Death among One's Age Group

Although a child may be on the surviving side of the coin of existence, that child experiences the loss of his or her siblings and age mates who are on the dying side. Figure 10.1 shows that for a group of infants born within the same time period, 30 percent perish in the first year of life and by the age of 10, 50 percent of the group are dead. One comes face to face with death early and frequently in this society. The Headlands' three children personally shared this experience. They were born in the Philippines in the 1960s and grew up in Agta camps. By the time they were out of their teens, the majority of their childhood friends were dead.

Death of Parents

The child who survives may also experience the loss of a father or mother at an early age. Table 10.1 estimates the percentage of Agta children who must confront this traumatic experience. The table was formulated from the database for Agta born after 1949, lived on the San Ildefonso Peninsula at some period in their life, were never members of the Acculturating population, and lost their father or mother before they themselves died. Fifteen years of age is taken as an estimate of when parental support is of lesser importance. Table 10.1 shows that approximately one-half of these survivors lost a parent before the age of 15.

Marriage and Death

For those who survive the childhood stages of the life cycle, the next stage is trial marriage and/or marriage. An early death of either a person or his or her spouse results in a brief union. Table 10.2 shows the outcomes of

Table 10.1. Percentage of Agta Children Who Survived One or Both Parents, 1950–94

Age of Survivor	Death of			
	Father		Mother	
	#	%	#	%
0	7	4.5	8	5.7
1	27	17.4	16	11.3
5	32	20.6	24	17.0
10	22	14.2	19	13.5
15+	67	43.2	74	52.5
Total	155	100	141	100

unions for those individuals who lived in the study population at some time, were married after 1949, and for whom dates of the beginning and end of their unions are available. For the first union of those less than age 20 years at the time of marriage, termination was primarily by divorce. Most of these were trial marriages, and the use of the term *divorce* for their termination is meant in the broad sense discussed in chapter 7. During an

Table 10.2. Percentage Marital Outcomes (Including Trial Marriages) for Individuals, 1950–94, by Age of Ego (the individual in focus)

First Marriages

Age Ego	N	Total %	Death Ego %	Death Spouse %	Divorce %	Ongoing %
<20	21	100	4.8	4.8	47.6	42.9
20	49	100	8.2	18.4	30.6	42.9
30	56	100	28.6	33.9	1.8	35.7
40	25	100	16.0	20.0	4.0	60.0
50	12	100	25.0	33.3	—	41.6
60	6	100	16.7	33.3	—	50.0
Total	169					

Second Marriages

Age Ego	N	Total %	Death Ego %	Death Spouse %	Divorce %	Ongoing %
<20	2	100	—	—	—	100.0
20	26	100	23.0	23.0	15.4	38.5
30	52	100	30.8	25.0	19.2	25.0
40	36	100	38.9	25.0	13.9	22.2
50	38	100	42.1	42.1	—	15.8
60	24	100	33.3	45.8	8.3	12.5
Total	178					

Third Marriages

Age Ego	N	Total %	Death Ego %	Death Spouse %	Divorce %	Ongoing %
<20	—	100	—	—	—	—
20	1	100	—	—	—	100.0
30	11	100	18.2	45.5	9.1	27.2
40	13	100	15.4	15.4	38.5	30.8
50	5	100	20.0	20.0	—	60.0
60	10	100	30.0	40.0	20.0	20.0
Total	40					

individual's 20s, the death of that person or his or her spouse is almost as important as divorce in terminating a first marriage. From age 30 on, mortality is by far the biggest terminator in first marriages. In second marriages death of an individual or spouse is by far the most important factor at all ages. In this high mortality population, mortality is more frequent than divorce in generating serial monogamy.

Parents and Survival of Offspring

Even if a couple can ward off their own deaths, they may still have to cope with the mortality of their offspring. As previously discussed, they lose 50 percent of their children within the first 10 years of the children's lives. If the couple survives to 45 years of age, they will have an average of seven to eight live births and lose three to four of these as young children. There is a notable exception to these averages, as shown in Figure 10.2. This married couple had 13 children, all living, in May 1995, 11 of whom appear in this photograph. The mother and father are fourth and ninth from the left.

SOME BEHAVIOR CHARACTERISTICS OF HIGH MORTALITY POPULATIONS

The experience of death and the outlook on death affect the psychology of a group. There have been no clinical studies of the San Ildefonso Agta to understand these effects. However, some behavioral characteristics have

Figure 10.2. An Agta woman, married in 1972 at age 14, has had 13 live births. All but the last were still alive in 1997. (J. Early, 1994.)

been observed that the coping with constantly high mortality may help to explain. This section speculates on the impact of mortality on these characteristics, none of which are specific to the Agta alone.

The Agta are unaccustomed to using foresight in planning for the future, compared with the use of Weberian "rationality" in the circumstances of industrial societies. Horace's Latin dictum could well be the Agta motto: *carpe diem*—seize the day, the present moment. With life so tenuous for both the individual and his or her family, the tendency is to focus on the present and give little thought to the future, which may never be for so many. This observation concurs with Woodburn's (1982b:207) findings:

> More generally, the idea of social continuity is not one which is stressed in immediate-return systems which are in so many respects strongly oriented to present activity. Economic activity is focused on immediate production and immediate consumption, and social activity in general is not burdened by substantial long-term concern, commitment, or planning. Consistent with this, the focus in death beliefs and practices is largely on immediate practicalities—the disposal of the corpse, the expression of personal grief and the grief of the wider community—rather than in precision for social replacement, reproduction, and long-term continuity.

This study would add that when high mortality is present, it is an additional factor for the immediacy characteristic of immediate-return systems.

In spite of this fatalistic attitude toward many things, the Agta give special attention and care to their infants and small children. They watch them, almost constantly hold them, and take safeguards to protect them from wind, rain, cold, night dew, and numerous spirit forces that may cause sickness or death. They protect them with various medicines—herbal, magical, and western. Parents are well aware of the high infant and child mortality (see LeVine 1977:21).

Another characteristic of the Agta is their usage of alcoholic drink, which leads to inebriation and sometimes death. This factor was discussed in chapter 8. It appears that the constant facing of one's own and others' mortality creates a stress that alcohol helps to relieve. Excessive usage of alcohol can arise from many causes, but in a society of high mortality, fear of death may be one of them.

The Agta, as in many small group societies, place a very high importance on kinship structures. Kinship is correlative with reciprocity, the logic by which goods and services are transferred within the group. The nuclear family is the basic kin unit. But due to the high mortality, the continuity of the same people constituting the nuclear family is tenuous. When death removes individuals with kin responsibilities, other kin must take over the responsibilities of the departed or else the group will not survive. This is one reason kinship assumes an importance in these cultures that it lacks in others.

STABILITY OF THE DEMOGRAPHIC STRUCTURE

Figure 6.1 shows the yearly change in the size of the San Ildefonso Agta population during the 44-year period. There are numerous fluctuations, especially the impact of the NPA guerrilla warfare in the middle 1970s. The overall trend is gradual increase so that there were 53 more people in the population in 1994 than in 1950, an average increase of slightly more than one person per year. This slight increase has taken place in spite of the dramatic ecological and social changes previously examined. The components of this increase can be examined to understand this stability. (Stability does not imply a "stable population" in the mathematical sense of the term as used by demographers.)

Stability of Fertility, Mortality, and Natural Increase

Figure 10.3 shows the curves of the crude birth and death rates for five-year periods. The relative flatness of the curves show their stability over the research years. Consequently, the resultant of the curves, the rate of natural increase in Figure 10.5, shows only a 20 per 1,000 population (2 percent) range for such a small population.

Stability of Net Migration

Figure 10.4 shows the crude rates of in-migration and out-migration for five-year periods. These curves show much more variability than those for the birth and death rates. In spite of the large fluctuations in each, their respective variations tend to cancel each other out over the entire 44-year research period so that their resultant, the net migration in Figure 10.5, is close to zero. The stability here is the cancelation effect. In terms of Figure 10.5, the total area for net migration below the zero line almost equals the total area above the line.

Stability of the Age-Sex Structure

The levels of the four demographic variables are partially determined by the age-sex structure of a population. In turn, the demographic variables themselves help determine the age-sex structure. Table 10.3 shows the age-sex structure of the Agta population for the three phases and the entire research period. If graphed, these distributions would yield the expected

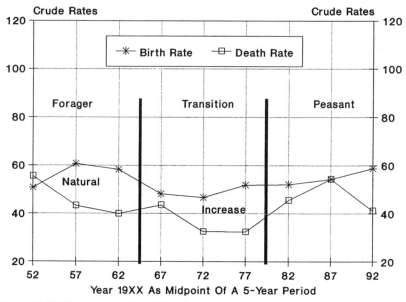

Figure 10.3. Five-year crude birth and death rates of Agta population, 1950–93.

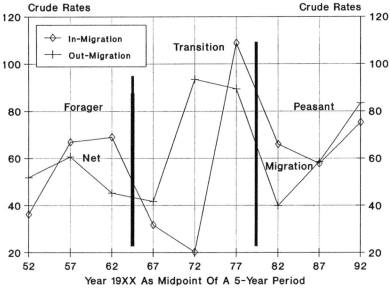

Figure 10.4. Five-year crude in-migration and out-migration rates, 1950–93.

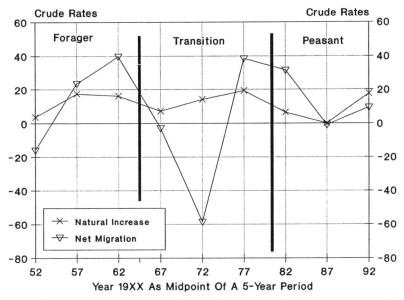

Figure 10.5. Five-year crude rates of natural increase and net migration, 1950–93.

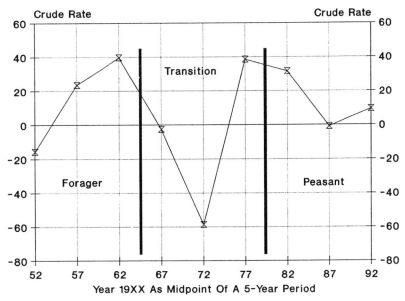

Figure 10.6. Five-year crude rates of total increase in Agta population, 1950–93.

pyramid which is the characteristic geometric form of populations with high fertility and high mortality. The sex ratio of the population with a slight male dominance has remained relatively constant over the research period. Migration in some other types of populations frequently changes the age-sex structures. But among the Agta, migration usually takes place in family groups so that the age-sex structure of migrating individuals is much the same as that of the rest of the population.

Stability of the Phases

The stability of all these factors means that the distinction of the three phases with their differing cultural and ecological conditions does not greatly discriminate the demographic rates, as shown in Table 10.4. The population increased by over 1 percent both in the Forager and Peasant Phases in spite of high mortality and high out-migration. The very high fertility overcame the high mortality, for an approximate 0.8 percent natural increase, and high in-migration overcame the high out-migration, for an approximate 0.5 percent net migration. The analysis shows that in spite of the very different cultural conditions of the San Ildefonso Agta in these two phases, their demographic structure remained much the same.

The Transition Phase was slightly different, because although there was lower mortality, there was an overall decrease of the population. Mortality was lower than in the other two phases, primarily because of the medical efforts of the missionaries and the expansion of Philippine health services. The lower mortality led to a 1.5 percent natural increase and would have been larger if fertility had not taken a small decline. However the large out-migration in 1975 resulting from NPA military activities took place during this phase leading to a -2.7 percent net migration. This more than offset the rise in natural increase from falling mortality so that the total decrease in this Transition Phase was -1.2 percent.

In the Forager and Peasant Phases the population grew by 43 and 41 people respectively, while in the Transition Phase it declined by 31 people. The net result was an increase of 53 people over the 44-year period, or a 0.6 percent annual rate of increase. This low rate of increase is quantitatively comparable to that of industrial societies, although it is due to an entirely different configuration of population dynamics. It is considerably lower than the 2.5 percent to 3 percent rate of growth of many agricultural societies. This stability of the population dynamics of the San Ildefonso Agta in spite of cultural change is contrary to what studies have found in some other foraging groups. This will be examined in chapter 12.

Table 10.3. Age-Sex Structure of Population, 1950–94, in Percentages by Phases

Age	Phase*			Total
	F	T	P	
Male	%	%	%	%
0	32	28	27	29
10	19	22	24	22
20	18	18	14	17
30	12	12	16	13
40	10	9	8	9
50	6	7	6	6
60	3	4	3	3
70	0	1	1	1
Total	100	100	100	100
% of Pop.	51.6	54.2	54.5	53.5
N	1,324	1,556	1,683	4,563
Female	%	%	%	%
0	18	19	21	19
10	22	21	15	19
20	19	16	17	18
30	14	16	16	15
40	10	13	9	11
50	7	6	8	7
60	2	2	4	3
70	0	0	1	0
Total	100	100	100	100
% of Pop.	48.4	45.8	45.5	46.5
N	1,242	1,317	1,404	3,963
Total	%	%	%	%
0	29	27	28	28
10	20	21	20	21
20	19	17	15	17
30	13	14	16	14
40	10	11	9	10
50	7	7	7	7
60	3	3	3	3
70	0	1	1	0
Total	100	100	100	100
N	2,566	2,873	3,087	8,526

* F = Forager Phase; T = Transition Phase; P = Peasant Phase

SUBSECTORS OF THE PENINSULA

This section examines the populations of the three subsectors of the peninsula that, when taken together, comprise the San Ildefonso Agta population. Its purpose is to find out if analysis at this more micro level will give additional insight into the population dynamics.

The Three Subsectors and Their Subgroups

The three subgroups of the San Ildefonso Agta will be designated by geographic terms for the subsectors of the peninsula in which they live—the Koso group, the Dinipan group, and the Lower Peninsula group (Map 4). The first two names are taken from the main rivers in the subsectors. In the Forager Phase the Agta resided in small, widely scattered camps throughout the forest and along the coast, with each camp typically consisting of three to seven kin-related nuclear families. If there was more than one camp group in the same river valley, there was a tendency for all the people in

Table 10.4. Size of Population in 1950 and 1994; Numbers and Crude Rates of the Demographic Variables, 1950–94, by Phases

1950		Phase*		Total	1994
	F	T	P		
Population 178		Increase 53			Population 231
Births	149	137	169	455	
Deaths	117	100	147	364	
Natural increase	32	37	22	91	
In-migration	145	140	200	485	
Out-migration	134	208	181	523	
Net migration	11	-68	19	-38	
Total increase	43	-31	41	53	
Crude rates					
Birth	58.0	47.7	54.7	53.3	
Death	45.6	34.8	47.6	42.7	
Natural increase	12.5	12.9	7.1	10.7	
In-migration	56.5	48.7	64.7	56.9	
Out-migration	52.2	72.4	58.6	61.3	
Net migration	4.3	-23.7	6.1	-4.5	
Total increase	16.7	-10.8	13.3	6.2	

* F = Forager Phase; T = Transition Phase; P = Peasant Phase

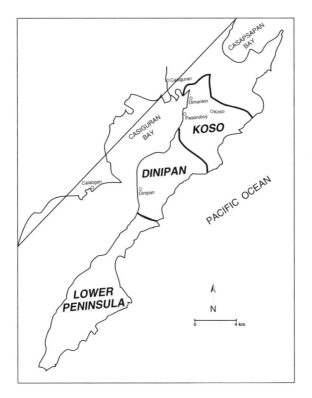

Map 4. Agta Subgroup Areas of the San Ildefonso Peninsula (based on PCGS, Solano, 1986).

these camps to be related. This relatedness prevented marriage within the river valley, so these groups tended to be exogamous. The marriage pattern was the basis for the research distinguishing the three groups. The changes since the Forager Phase have altered this pattern. There has been migration between the three subgroups. Related Agta families no longer tend to reside together but are interspersed among lowlanders with whom they have daily political and economic interaction. For administrative purposes, the local and national governments recognize the two northern subsectors, Koso and Dinipan, as a single subjurisdiction of the municipality of Casiguran and the Lower Peninsula as another subjurisdiction.

The Koso group lives on the northern part of the peninsula, and its northern boundary divides the whole peninsula from the rest of Casiguran. This boundary is a line that begins at the mouth of the Kasalogan River (16° 16′

N by 122° 8′ E), runs north-northeast for about 1,200 meters, then turns and runs in a straight line east-southeast to the east coast of the peninsula, ending at 16° 16′ N by 122° 15′ E. The southern boundary dividing Koso from Dinipan begins at the mouth of the Kataguman River and follows the river in a southeasterly direction to the east coast at Lublub Point. The Dinipan subsector is the central part of the peninsula and is divided from the lower peninsula by the Didumayan River. The lower peninsula is the southern portion of the study area and is surrounded by the Pacific Ocean and the Casiguran Bay.

Characteristics of the Three Subsectors

In 1960 Koso had more full-growth forest than the other two peninsula subsectors. As a consequence, there was more emphasis on hunting in this subsector. Its rivers are larger than in other parts of the peninsula, and river fishing and shell fishing in the mangrove swamps were more important than marine fishing. Koso was the first part of the peninsula to be penetrated by loggers, which led to an earlier depletion of the forest canopy and game than in the other two subsectors. Both river and marine fish have also been badly depleted, and much of the coral reef has been destroyed from dynamite fishing. The loss of these resources has led to a weakening of the ahibay system. The lowlander in-migration began in the 1920s. It was slow until the late 1970s, when the government opened Koso for homesteading. Of the three subsectors, Koso was first to lose its semi-isolation and consequently has suffered earlier environmental degradation.

In the 1960s the Dinipan subsector was less forested, its water courses smaller, and hunting was less important than in Koso. The Agta in this subsector placed more emphasis on swidden cultivation. Both river and marine fishing were important activities until the 1970s. Because this subsector was more isolated than Koso from the rest of Casiguran, the lowlander invasion came later. The ahibay system in Dinipan has weakened in the Peasant Phase but remains stronger than in Koso.

The lower peninsula is the least forested subsector and has only a few small streams. Agta camps tend to be more scattered than in the other two subsectors. Historically there was some hunting, but it was of less importance than in Koso. The subsistence orientation of this subsector has always been marine fishing. Much of the coral reef surrounding it is still alive and the fishing still adequate. The ahibay system is still important here and retains more of the traditional characteristics. The first two lowlander families migrated to the Lower Peninsula in 1940 and 1942, and the heavy influx of lowlander homesteaders began in the late 1980s.

THE POPULATION DYNAMICS OF THE THREE SUBGROUPS

Figure 10.7 and Table 10.5 show the distribution of the Agta population of the peninsula among the three subgroups. From 1950 to 1994 Koso has decreased its share of the Agta population by 13 percent, while Dinipan increased its share by 2 percent and the lower peninsula by 11 percent. Table 10.6 examines the components of increase or decrease for each subgroup for the entire research period. Figures 10.8 to 10.14 are graphs of the rates of these components for five-year segments. An inspection of Figure 10.8 shows the crude birth rates approximate one another, with the exception of Koso. In Figure 10.9 the crude death rates show more variation than the birth rates, as expected. Koso again has several larger deviations than the rates for the other two subsectors. Figure 10.12 shows the rates of natural increase, the resultant of the birth and death rates. For such small groups, the range of variation is slight. Figure 10.10 shows relative uniformity of in-migration among the subsectors. Dinipan is the exception in the late 1950s and early 1960s. There were food shortages in the Palanan Valley 90 kilometers (56 miles) to the north in this period, and a

Table 10.5. Area, Population, and Density of the Three Agta Subgroups of the San Ildefonso Peninsula

Population	Peninsula	Koso	Dinipan	Lower Pen.
1950	178	52	74	52
1994	231	37	100	94
d*	53	-15	26	42
1950	100%	29.2%	41.6%	29.2%
1994	100%	16.0%	43.3%	40.7%
d	—	-13.2%	1.7%	11.4%
Area
km²	108.8	41.1	21.5	46.2
mi²	42.0	15.9	8.3	17.8
Density per sq km
Agta Population
1950	1.6	1.3	3.4	1.1
1994	2.1	0.9	4.7	2.0
Lowlander Population
1942	2.0	—	—	—
1994	21.2	—20.8—		21.7

* d = difference (between the numbers in the two previous rows)
Source: Areas by planimeter courtesy of Prof. Dayle Clark, University of Texas, Arlington

number of Palanan Agta in-migrated to the Dinipan subsector. Also, a number returned to Dinipan after working at the magnesium mines at Dinapigui. Figure 10.11 showing out-migration indicates relative uniformity except for Dinipan and Koso during the last period. Some of this out-migration from Dinipan represents families going to the mines. Figure 10.13 shows the rate of net migration, the resultant of the rates of in-migration and out-migration. Again the profiles of the three subgroups tend to approximate one another. Figure 10.14 shows the total rates of increase, the net result of the rates in Figures 10.8 to 10.13. These rates of increase result in the absolute changes of the population in Figure 10.7.

Koso

Koso's profiles are somewhat different from the other two subsectors. Figure 10.8 shows that the low natural increase in 1985–89 resulted from low fertility. The rate may be due to the small size of the Koso Agta subpopulation, since it represents nine births for 46 women of reproductive age. There was no radical change in the age-sex structure in this period. However, the lower fertility in this period was a continuation of the trend in Koso

Table 10.6. Crude Rates of the Three Agta Subgroups: Koso, Dinipan, and Lower Peninsula, 1950–94

Crude Rates	Peninsula	Koso	Dinipan	Lower Pen.
Birth	53.3	49.1	51.0	59.8
Death	42.7	42.3	41.3	44.6
Natural increase	10.7	6.9	9.7	14.9
In-migration	56.9	53.3	88.1	55.0
Out-migration	61.3	67.0	90.9	54.3
Net migration	-4.5	-13.8	-2.8	.7
Total increase	6.2	-6.9	6.9	15.7
N				
Births	455	107	184	164
Deaths	364	92	149	123
Natural increase	91	15	35	41
In-migration	485*	116	318	151
Out-migration	523*	146	328	149
Net migration	-38	-30	-10	2
Total increase	53	-15	25	43

* The sum of the subgroups does not equal the study population because of internal migration between the subgroups, which is not migration for the study population. However, this movement cancels out, so the net migration remains the same.

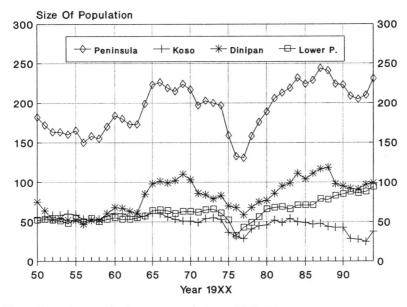

Figure 10.7. Agta and subgroup populations, 1950–94.

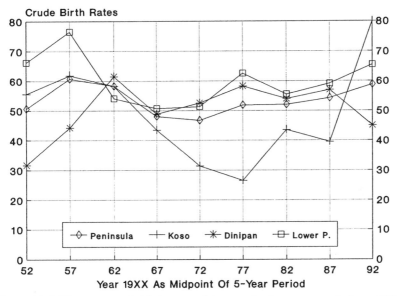

Figure 10.8. Five-year crude birth rates of Agta and subgroup populations, 1950–93.

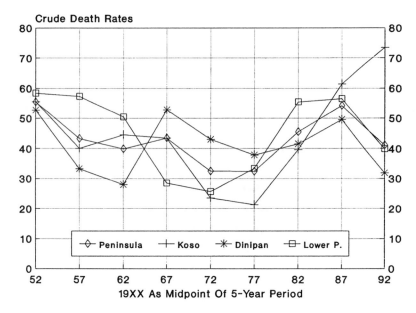

Figure 10.9. Five-year crude death rates of Agta and subgroup populations, 1950–93.

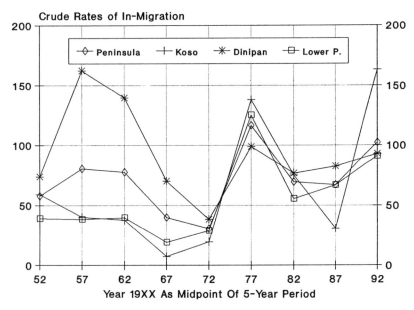

Figure 10.10. Crude rates of in-migration of Agta and subgroup populations, 1950–93.

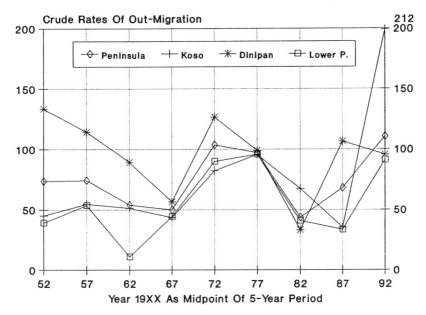

Figure 10.11. Crude rates of out-migration of Agta and subgroup populations, 1950–93.

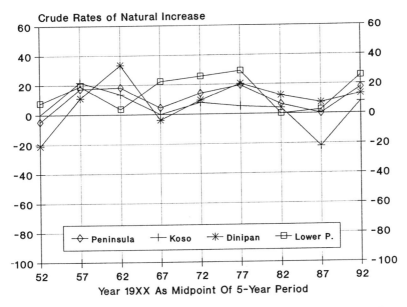

Figure 10.12. Crude rates of natural increase of Agta and subgroup populations, 1950–93.

toward lower fertility that began around 1970. Figure 10.8 shows that the drop in fertility in the Transition Phase for the whole San Ildefonso Agta population, discussed in chapter 7, was due entirely to the low number of births during that phase in the Koso subpopulation. The reason for the decline in births in Koso during that phase is unknown.

Figure 10.9 helps explain the decline in mortality for the whole population noted in chapter 8. The Koso rates were the lowest for most of the Transition Phase. This trend was attributable to the paramedical programs of the missionaries and the increase of medical facilities in Casiguran. The Headlands lived in Koso during this period, and their interaction with the Koso Agta was greater than with other groups. Koso was also the closest subgroup to Casiguran where medical facilities were located. The Koso subgroup were thus the first to recognize the efficacy of western medicines.

Figure 10.10 shows high in-migration to Koso during the 1990s, but Figure 10.11 shows that this was offset by an extraordinary amount of out-migration, which contributed to the decline of the Koso population. One reason for this out-migration is the high mortality in the last five-year period. There were 36 out-migrations, including six children who left to live

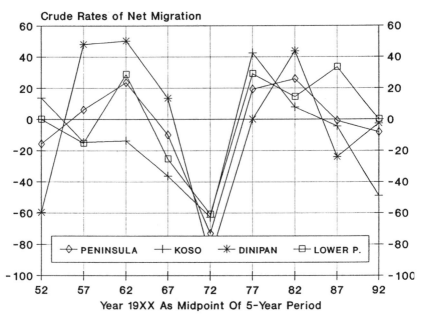

Figure 10.13. Crude rates of net migration of Agta and subgroup populations, 1950–93.

with other kin because of the deaths of one or both parents, and two widows who returned to their own kin groups because of their husbands' deaths.

Homogeneity of the Subgroup Data

The demographic events of the subgroups involve relatively small numbers. In spite of potential volatility from the small numbers, this section has shown that the population dynamics of the three subgroups are quite similar. In addition, the data are homogeneous in spite of the differences between the subgroups and in spite of the substantial changes of social and physical ecology that have come over the area. The only large deviation was the impact of the NPA and government warfare in 1973–76, but the effect was not permanent. In the midst of all these changes, the San Ildefonso Agta have maintained themselves by a stability resulting from a precarious balance in which very high fertility has overcome high mortality and large in-migration has usually offset high out-migration.

FACTORS THREATENING THE STABILITY OF THE AGTA POPULATION

As shown in Figures 10.8 and 10.9, the San Ildefonso Agta maintain their population in the face of high mortality primarily by their high fertility,

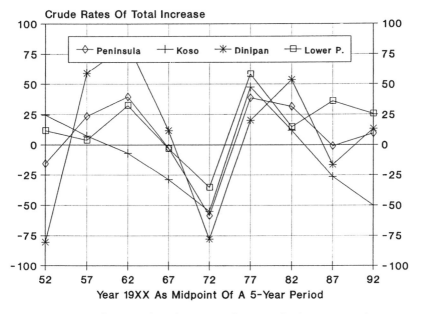

Figure 10.14. Crude rates of total increase of Agta and subgroup populations.

whose crude rates usually reach the mid 50s and whose total fertility is seven to eight live births for women who survive their whole reproductive period. Any altering of this delicate balance, such as an increase in the use of substitutes for breast-feeding, would be highly threatening to the group. This effect was examined in chapter 8. If contact with Filipino peasantry leads to abandoning breast-feeding and the substitution and mishandling of sugar water, the infant and child mortality rates could increase to a level where the high fertility would be incapable of overcoming them. (Fertility also would probably increase under these conditions, as the increase in infant and child mortality would shorten the birth intervals. But the increase in mortality would probably outstrip the increased level of fertility.)

Lowering fertility without lowering mortality would be highly threatening. As the Agta increase interaction with Filipino peasantry, such changes could take place. If Agta seek education, or if the increasing numbers of Agta girls moving to Manila continues, it will probably lead to later marriage for the girls and consequent delay in initiating childbearing. Mortality may also fall in this situation, but if it does not, the lower fertility would mean decline of the population.

The San Ildefonso Agta population has also been able to maintain itself by high levels of in-migration that usually offset high out-migration. Any increase in out-migration without any accompanying increase of in-migration would be a threat. Such an increase in out-migration is a possibility. The peninsula has been through a change of demographic density in the 44 years covered by this research. In 1950 there were 178 Agta and 250 Filipinos along with the game on the rain-forested peninsula. In 1994 the Agta of the San Ildefonso Peninsula exist as a small minority of 231 people living in the midst of about 2,500 Filipinos (Table 3.2) and a completely changed physical ecology. The San Ildefonso Agta are now in almost complete economic dependence on the Filipino lowlanders. This dependence and closer interaction between Filipino and Agta have led to interethnic hypergynous marriages, causing cultural out-migration of Agta from their traditional population. This interaction is the topic of the next chapter.

Part V

Acculturation

Chapter 11

The Acculturating Population

In the last half-century the Casiguran Agta have changed from foragers sharing a remote region with a sparse lowlander population to a small peasant minority in the midst of a sizable and fast-growing lowlander population dominated by land-seeking Filipino in-migrants. In terms of rural stratification, many Agta are the lowest class—land-short or landless peasants without tenancy rights, and within this class they comprise a subordinate ethnic group. They are in daily and close interaction with lowlanders, some of whom are also landless peasants. There have been conflicts between individual Agta and lowlanders. However, relationships between the two groups are not always hostile.

Agta-lowlander marriages have occurred. As would be expected in a socially subordinate group, most cases are Agta women marrying lowlander men (hypergyny). At the beginning of the Forager Phase in 1950 all San Ildefonso Agta married women had Agta men as spouses. During the Forager Phase, 97 percent of marriages entered into by Agta women were to other Agta; in the Transition Phase it was 81 percent, and in the Peasant Phase it dropped to 64 percent. In some of these marriages, the Agta woman had access to the land where her group traditionally lived. When hypergynous marriages have occurred, the lowlander husband also gains access to this land by the marriage. Land acquisition is not the only reason for such marriages, but in many cases it is the main attraction. This chapter investigates the impact of these interethnic marriages on the population structure of the peninsula.

CULTURAL CHANGE

For the analysis, we have defined a third distinct population living on the San Ildefonso Peninsula in addition to the Agta and lowlander populations: *the Acculturating population.* This label considers this population from the viewpoint of the effects of interethnic marriage on the Agta member and her descendants. In these marriages the male of the household is a Filipino socialized in rural Filipino culture and speaking one of the major Philippines languages, usually Tagalog or Bikolano. The couple live on a lowlander farm or in a lowlander hamlet on the peninsula. An Agta woman in such a union becomes immersed in her husband's way of life and loosens her ties to Agta ways. This is the sense of the label "Acculturating" as it is used here. Children of these marriages are more culturally Filipino than Agta. For the descending generations, the criterion is that a person descended from such a household retains these Filipino characteristics. Figure 11.1 shows an interethnic couple with some of their children. The photograph at the beginning of the chapter is a close-up picture of an Agta wife and Filipino husband.

Figure 11.1. An Acculturating family standing in front of lowlander style house. A closeup of the husband and wife appears in the introductory photo to this chapter. (J. Early, 1994.)

Demographically, an interethnic marriage on the peninsula involves a cultural out-migration from both the Agta and lowlander populations and two cultural in-migrations into the Acculturating population. (There is only one exception: In 1987 a lowlander man married an Agta woman and took up residence in an Agta settlement. He is the only lowlander considered a member, as an in-migrant, of the San Ildefonso Agta population.) If an interethnic marriage is terminated, more cultural migration may take place. If a lowlander father should die while his child is young and the child is brought up by the Agta widow and her Agta kin group, then the widow and her child are considered as culturally out-migrating from the Acculturating population and in-migrating back into the Agta population.

There have been 186 people who have been members of the Acculturating population at some time since 1951. Of the 2,832 people living on the San Ildefonso Peninsula at the beginning of 1994, 101 (3.6 percent) were members of this population, while the Agta population was 231 (8.2 percent) and the lowlander population was about 2,500 (88.3 percent). (See Table 11.1, Figure 11.2, and Table 3.2).

GROWTH OF THE ACCULTURATING POPULATION

The Acculturating population began with an interethnic marriage in 1951. (There were a few children of interethnic parents on the peninsula before 1951 but none from marital unions and all were brought up as Agta.) The lowlander husband of that union died in 1955. His Agta widow then married an Agta and with the surviving child returned to the San Ildefonso Agta population. But before that first husband's death, another interethnic marriage took place in 1954 and was still intact in 1994. This was the only Acculturating family until 1966. There were two more interethnic marriages during the late 1960s and four in the 1970s along with one in-migrating family. Therefore, the Acculturating population was small during the Forager and Transition Phases (see Figure 11.2). Marriages of Agta women to lowlanders are primarily a phenomenon of the Peasant Phase.

Table 11.1. Growth of the Acculturating Population, 1952–94

Year (Jan. 1)	Population	Yearly % Inc.	Phase*
1952	4	—	F
1965	8	5.5%	T
1980	34	10.1%	P
1994	101	8.0%	P

* F = Forager Phase; T = Transition Phase; P = Peasant Phase

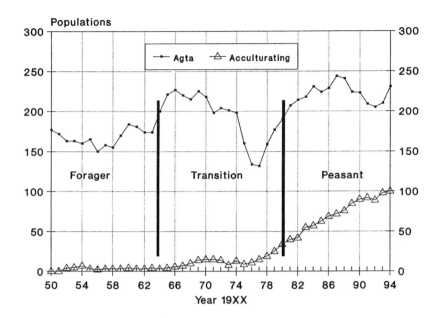

Figure 11.2. Size of Agta and Acculturating populations, 1950–94.

ENTRY INTO THE ACCULTURATING POPULATION

Lowlanders

The first part of Table 11.2 details the various ways in which 34 lowlanders have become members of the Acculturating population: 24 were married on the peninsula to San Ildefonso Agta; 5 married Agta outside the peninsula and later in-migrated along with their Agta spouses and their 10 children; 5 married second-generation members of the Acculturating population. Second-generation members are the offspring of interethnic marriages or are Agta children who were brought into the Acculturating population by their mothers at the time of marriage to lowlanders (who become the children's stepfathers). Table 11.2 shows only 1 female lowlander married to an Agta male. However, of the 5 lowlanders married to second-generation members of the Acculturating population, 3 are female.

Agta

Table 11.2 also shows that on the peninsula 20 Agta women have married lowlanders. (This number does not equal the number of lowlanders just mentioned because some Agta women have been married to more than one lowlander and vice versa.) They brought with them 11 children from previous marriages to Agta men; those children were then raised with the lowlander stepfathers. There were 7 Agta women from outside the peninsula who married their lowlander husbands before in-migrating to the peninsula. Three other Agta became members of the Acculturating population when they married second-generation members of that population. Two already lived on the peninsula and thus made a cultural migration. The third woman married a lowlander away from the peninsula and later moved onto the peninsula with him, in-migrating physically and culturally to the acculturating population.

Table 11.3 traces the marital histories of the 29 Agta women who entered into interethnic marriages. It shows that only 45 percent (13/29) of

Table 11.2. Sources of the Acculturating Population

	Total	Male	Female	Prior Children*			
				T	M	F	U
I. Lowlanders Married to							
A. Agta							
1. On peninsula	24	24	—				
2. Outside, in-migrated	5	4	1	10	6	4	—
B. Acculturating population	5	2	3				
II. San Ildefonso Agta Married to							
1. Lowlanders	20	—	20	11	7	3	1
2. Acculturating population	2	1	1				
III. In-migrating Agta Married to							
1. Lowlanders	8	1	7				
2. Acculturating population	1	—	1				
Totals	65	32	33	21	13	7	1
IV. Second Generation	101						
Not born in	21	confer above					
Born in	80						
V. Third Generation							
Born in	20						
Total	186						

* T = Total, M = Male, F = Female, U = Sex Unknown

Table 11.3. Marital Histories of Agta Women in Interethnic Marriages

Agta Women in Interethnic Marriages	Marriages							
	First to	Result	Second to	Result	Third to	Result	Fourth to	Result
	Agta 13	D.Sp. 8	Agta 1	D.Sp. 1	Lowl. 1	C. 1		
			Lowl. 7	D.Sp. 1	Agta 1	D.Sp. 1		
				Div. 1	Lowl. 1	C. 1		
				C. 5				
		Div. 5	Agta 1	Div. 1	Agta 1	Div. 1	Lowl. 1	Div. 1
29 Agta Women			Lowl. 4	Div. 3	Lowl. 2	Div. 2	Lowl. 2	C. 2
				C. 1				
	Lowl. 13	D.Sp. 2						
		D.Ego 1						
		Div. 2	Lowl. 2	Div. 2	Lowl. 1	Div. 1		
		C. 8	Lowl. 2	Div. 1				
				C. 1				
	Acc.P. 3	Div. 1	Lowl. 1	Div. 1				
		C. 2						

Abbreviations:
C.—marriage continues; D.Ego—death of ego; Div.—divorce; D.Sp.—death of spouse; Lowl.—lowlander; Ego—person in focus; Acc.P.—Acculturating population

the first marriages of these women were with lowlanders. Ten percent (3/
29) of the Agta women entered the Acculturating population by marrying
a second-generation member of that population. Another 13 (45 percent)
had entered into a first marriage with another Agta.

Of the 13 first marriages to lowlanders, 8 were ongoing in 1994, 1 had
been terminated by the death of the Agta wife, 2 by death of the lowlander
husbands, and 2 by divorce. Among the Agta whose first marriages were
to Agta spouses, 5 were terminated by divorce and 8 by death of the Agta
spouse. All but 2 of these 13 women entered into second marriages with
lowlanders. Thus the high Agta mortality becomes a contributing factor to
the growth of the Acculturating population. It increases the possibility of
Agta widows marrying lowlander men and in some cases bringing Agta
children with them.

Second- and Third-Generation Members of the Acculturating Population

While interethnic marriages continue, the Acculturating population has been
in existence long enough so that 56 percent of those who have belonged to
it were born into it as second- or third-generation members (Table 11.2).
The second-generation members consist of 80 individuals born into the
Acculturating population and 21 others who were either physical in-migrants to the peninsula or cultural in-migrants with their Agta mothers
after those mothers' previous Agta marriages. By 1994 there had been 10
marriages of second-generation members: 5 to lowlanders, 5 to Agta, and
none to each other. If this trend continues, it will accelerate the growth of
the Acculturating population relative to the other two populations. There
are 20 children in the third generation of the Acculturating population as
of 1994. All were too young for marriage at that time.

MAIDS AND INTERETHNIC MARRIAGE

In chapter 9, about migration, it was seen that Agta women working as
domestic servants in lowlander towns have been a source of Agta out-migration. Young girls become live-in servants in Filipino towns for several
years. Most eventually return, but their work experience is an acculturating experience. They learn a regional Philippine language and gain familiarity with Filipino customs. This facilitates marriage to lowlanders. During the 44-year research period, 33 San Ildefonso Agta girls worked at one
time as maids. Ten remained unmarried as of January 1994, as they were
in their teens or early 20s. Of the remaining 23, about half had become

members of the Acculturating population by 1994, 10 by marriage with lowlanders and 1 by marriage with a member of the Acculturating population. The ex-maids comprise 38 percent of the Agta women who have entered the Acculturating population by marriage. The custom of working as domestic servants decreases Agta population growth in two ways: For those who marry Agta, it delays the age at marriage and first birth; for those who marry lowlanders, they culturally out-migrate from the Agta population.

Lowlanders in Interethnic Marriages

There is information about the place of origin for 15 of the 34 lowlanders in the interethnic marriages. Six of them are from the municipality of Casiguran (5 from the town, 1 from the peninsula). One is from the adjoining municipality of Dinalongan. Three are Ilokanos from the province of Cagayan. Two are from the Bicol area of southern Luzon, and 2 are from the Visayan Islands in the middle of the Philippine archipelago.

The Population Dynamics of the Acculturating Population

Figure 11.2 outlines the growth of the Acculturating population. In 1951 it consisted of a single family of four. In 1994 it consisted of 101 people. Table 11.4 indicates a very high rate of total increase, over 8 percent per year. Fertility is very high, since this is a young population—more than 75 percent are below 30 years of age. Mortality is moderate, so the natural increase of over 4 percent is dominated by the high fertility. In-migration is substantially higher than out-migration, giving a high rate of net migration. Each of these sources of population growth can be examined and compared with the San Ildefonso Agta population.

Fertility

Table 11.5 shows the total fertility rate of the Acculturating population, which expresses the average number of live births for women who survive the reproductive period. It is extremely high: 9.2. For the Agta population in the Peasant Phase the average was also high: 7.6. One reason for the extremely high average of the Acculturating population can be seen by comparing the age-specific rates with those of the Agta. The large differences are in the 15–19 and 20–24 age categories. A methodological factor is partly responsible for these large differences. In cases of first marriages, a young female Agta does not become a member of the Acculturating population until she marries a lowlander. Methodologically this means the person-

Table 11.4. Demographic Factors in the Growth of the Acculturating Population, 1951–94

1951 (Jan. 1)	Crude Rates			1994 (Jan. 1)
Population 0				Population 101
	Total increase	80.4	(101)	
	Birth	78.0	(97)	
	Death	33.8	(42)	
	Natural increase	44.2	(55)	
	In-migration	85.2	(107)	
	Out-migration	49.0	(61)	
	Net migration	36.2	(46)	

Age	%M	%F
0	24.2	21.8
15	14.2	15.6
30	10.2	8.6
45	3.2	2.2
Total	51.8	48.2

N = 1,244 person-years

years of these young women while not childbearing—because they are not yet married—are part of the person-years of the Agta population. They increase the denominators of the fertility rates, thereby lowering the rates. Conversely they tend to increase the fertility rates of the Acculturating population because their person-years only start with marriage, which is closely followed by childbearing.

Table 11.5 indicates another reason for the extremely high fertility rate of the Acculturating population. The births of the Acculturating population are more closely spaced than those of the Agta: two and a half years compared with two years and ten months, a difference of four months between each birth. This difference may be due to shorter periods of breastfeeding in the Acculturating population.

Mortality

As indicated in Table 11.6, the level of mortality of the Acculturating population is substantially lower than that of the Agta population. The life expectancy at birth for the Acculturating population is 15 years greater, 37 versus 22 years for the Agta in the Peasant Phase. Child mortality is much lower than that of the Agta. More than 50 percent of an Agta cohort is dead by five years of age, whereas death of 50 percent of an Acculturating cohort does not occur until about 30 years of age. There are no mortality indicators of the previously discussed synergism between malnutrition and infectious disease. The child mortality rate of the Acculturating population is less than its infant mortality rate. Within the infant mortality rate there are only five cases with sufficiently accurate data about age of death, but all five are neonatal deaths. While the number is small, the presence of the

Table 11.5. Fertility Rates for the Acculturating Population and the Peasant Phase of the Agta Population

Level	Rate	Acculturating	Peasant Agta
Avg. female	Total fertility	9.2	7.6
Age groups	Age specific		
	15	337	161
	20	380	283
	25	270	341
	30	409	320
	35	301	294
	40	146	130
	Population female 15–44		
	as % of total population	24.2%	20.3%
Population	Crude birth	78.0	54.7
	Number of births		
	Male	48	84
	Female	47	85
	Unknown sex	2	0
	Total	97	169
	Birth intervals		
	All intervals	2.5 years	2.84
	Regular	2.7	3.1
	With neonatal death	1.4	2.4
	With death to 2.25 years	1.7	2.5
	Number of intervals	24	134
	Regular	20	84
	With neonatal death	2	14
	With death to 2.25 years	2	34
Female average age at first marriage		18.4	18.3
	N	20	35

synergism would have been indicated if there had been more infant deaths after six months.

In-Migration

The rates of migration for the Acculturating population in Table 11.4 reflect both cultural and physical migration. Interethnic marriages mean cultural migration for both parties into the Acculturating population along with any children from previous marriages. The rate for all in-migration is 85.2. Over half of this rate, 43.7, is due to marriage into the population.

Out-migration

Members of the Acculturating population out-migrate from the peninsula for many of the same reasons as the Agta, especially for economic or familial reasons. There have been nine cases of cultural out-migration involving a return to Agta or lowlander populations. One lowlander left because of divorce. One second-generation female left to marry an Agta and live in an Agta family group. The remainder of the out-migrations are Agta women and their children who, upon the death of or divorce from a lowlander spouse, returned to the Agta population by remarriage to an Agta or by simply going back to live with Agta kin.

Net Migration

Table 11.4 shows that the rate of in-migration is the highest rate of the four components of population growth (births, deaths, in-migration, and out-migration), even higher than the high level of fertility. Consequently net migration is also high, in spite of substantial out-migration.

Graphic Overview

Figures 11.3, 11.4, 11.5, and 11.6 present in graphic form for five-year periods the crude rates of the demographic variables discussed in this section and their interaction in determining the growth of the Acculturating population. The rates before 1980 are composed of very small numbers giving rise to some unusual values.

THE IMPACT OF THE ACCULTURATING POPULATION ON THE AGTA POPULATION

What Might Have Been

At the beginning of 1994 there were 16 living adult female members of the Acculturating population who had been members of the San Ildefonso Agta population before their marriage. They brought with them six surviving

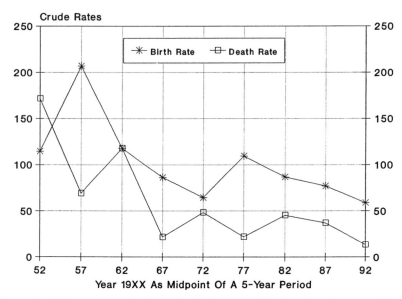

Figure 11.3. Five-year crude birth and death rates of the Acculturating population, 1952–93.

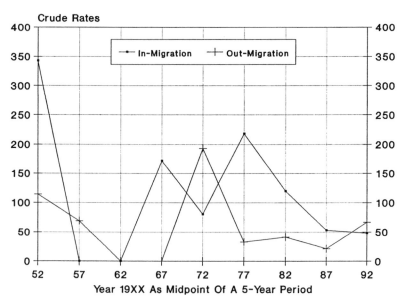

Figure 11.4. Five-year crude rates of in-migration and out-migration of the Acculturating population, 1952–93.

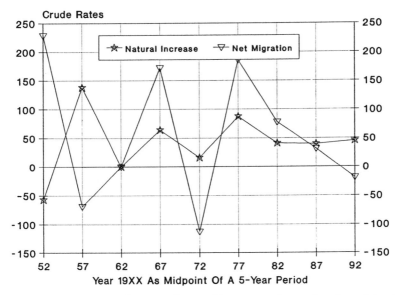

Figure 11.5. Five-year crude rates of natural increase and net migration of the Acculturating population, 1952–93.

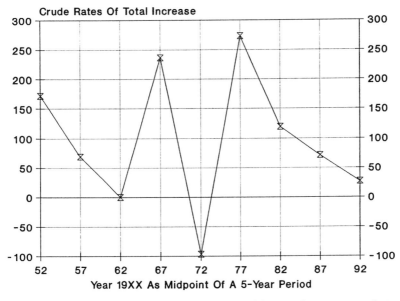

Figure 11.6. Five-year crude rates of total increase of the Acculturating population, 1952–93.

children from their previous marriages to Agta men. These 22 members of the Acculturating population in 1994 would still have been members of the Agta population if the interethnic marriages had not taken place, which in turn means there would have been 253 members of the Agta population in 1994 instead of the 231. The rate of total increase (r) of the San Ildefonso Agta population since 1950 would have been 0.8 percent instead of 0.6 percent.

Marriage Partners for Agta Men

Does the loss of San Ildefonso Agta women in interethnic marriages mean the male members of the population are having difficulty finding marriage partners? It does not appear so. In 1994 there were 23 San Ildefonso Agta

Table 11.6. Mortality Rates for the Acculturating Population and the Peasant Phase of the Agta Population

Level	Rate	Acculturating	Peasant Agta
Avg. person	Life expectancy at birth (years)	36.9	22.2
Age groups	q Probability dying during age interval		
	0	227	266
	1	181	342
	5	81	99
	10	103	47
	20	35	81
	30	157	231
	40	—	311
	50	—	443
	60	—	641
	70+	1,000	1,000
	N	42	147
	l—% Surviving to Age	%	%
	0	100	100
	1	77	73
	5	63	48
	10	58	44
	30	50	38
	50	43	20
Population	Crude Death	33.8	47.6

men ages 20 to 29. Of these, 70 percent (16) were already married. Of the seven unmarried, four were 22 years of age and one 23 years. It is not unusual for Agta men to be still single at these ages. The remaining two were 26 and 27 years old. Because of the proscription against marrying anyone who is called by a consanguineal or affinal kin term, the Agta men of the peninsula frequently have gone to other Agta areas seeking marriage partners and continue to do so. For this reason, there has been a ratio of three female Agta in-migrants for marriage to San Ildefonso Agta men for every male in-migrant for marriage during the Peasant Phase.

THE FUTURE OF THE SAN ILDEFONSO AGTA AND ACCULTURATING POPULATIONS

Demographic Equality

The biggest threat to the maintenance of the San Ildefonso Agta population is hypergyny—increased out-migration of women owing to increased marriages with lowlander men. As lowlanders continue their in-migration to the peninsula and the Agta become an even smaller minority in their midst, the rate of interethnic marriages will probably increase. Such marriages are a double-edged demographic sword because they accelerate the relative growth of one population and the decline of the other. A single marriage not only takes a person away from the Agta population but also adds one to the Acculturating population. The result is a net difference of two between the populations, while one event of fertility or mortality in a population results in a difference of only one (see Early 1982:88–95).

In 1994 the Agta population was over twice as large as the Acculturating population, 231 to 101. But the growth rate of the Agta population was low, 1.33 percent per year in the Peasant Phase, while the Acculturating population was exploding at a rate of 8.04 percent per year. To give some idea of how quickly the Acculturating population is overtaking the Agta population, Table 11.7 is a hypothetical projection (not a prediction) based on the assumption that the growth rates of the Peasant Phase for both populations will remain the same. Under these conditions, the two populations would be equal in size around the year 2007. It is more likely to be sooner because the growth rate of the Acculturating population will probably accelerate.

Outgroup marriage is aided by the fact that the Agta are not highly defensive of their cultural ways. They know their subordinate position and seek to survive in it with little thought of preserving their traditional culture. This concurs with Woodburn's (1980:106) observation that people in

Table 11.7. Size of the Agta and Acculturating Populations: 1950–94 Actual, 2000–2030 Projected

Year	Population		Ratio	
	Agta	Acculturating	Agta:Accult.	Accult.:Agta
1950	177	0	—	
1960	184	4	46.1:1	
1970	218	15	14.5:1	
1980	190	34	5.6:1	
1990	223	90	2.5:1	
1994	231	101	2.3:1	
2000	250	161	1.6:1	
2006	271	257	1.1:1	
2007	274	278	1:1	1:1
2010	285	351		1.2:1
2015	305	517		1.7:1
2020	326	763		2.3:1
2030	372	1,660		4.5:1

simple foraging societies do not "seem to value their own culture and institutions very highly and may . . . not be accustomed to formulating what their custom is or what it ought to be."

The Future of the Agta

Will the San Ildefonso Agta survive their change from mobile foragers of the rain forest to peasants within the Philippine nation? The question needs some distinctions. Will their culture survive in its traditional form, as when they were foragers? As seen in the previous chapters, there have already been significant changes. It may survive in a modified form as an ethnic subgroup within Filipino culture. But this premise depends on a more basic question. Are the Agta people themselves going to die out? Unless their high mortality continues to be offset by their high fertility, extinction is always a possibility. Physical migration in both directions is highly volatile and the excess of in-migration over out-migration is problematical.

But there is another alternative to dying out. The Acculturating population as it has been defined in this research is probably a temporary population, a transient stage. It provides a cultural matrix in which the mixed progeny learn the national culture in a family setting. This environment leads to assimilation of the Agta into the Filipino multiracial population, as has happened among Negritos in other parts of the country (e.g., Eder 1987). Any San Ildefonso Agta who continue to be unassimilated will remain landless peasants whose role will be to provide cheap labor for local farmers. This situation may lead to further deterioration of health conditions, to an increase in mortality, and perhaps to extinction.

Part VI

A Comparative Perspective

Chapter 12

The Population Dynamics of Foraging Societies

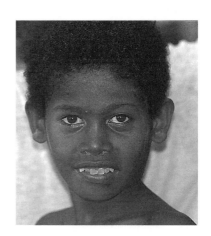

This research has examined the population dynamics of the San Ildefonso Agta over a 44-year period. Several questions arise. How typical is this case of the Agta compared with other foraging societies? What further light can the literature about foraging societies throw on this case? What contribution can this Agta case make to that literature? This chapter will take a comparative perspective and examine these questions.

FORAGING SOCIETIES

The current literature about foraging societies (Ingold et al. 1988, Lee 1992, Burch and Ellanna 1994, Spielmann and Eder 1994, Kelly 1995, Kent 1996) is divided about what constitutes a foraging society, with some writers even questioning whether such a social type exists. Regardless of whether a distinctive social type exists or not, this Agta case study takes the position that these societies have a distinctive type of subsistence activity and style of life. Whether or not this implies a distinctive demography is one of the questions being examined.

Because of the diversity among foraging societies (Kelly 1995; Kent 1996), the category needs some classificatory distinctions. For the purposes here, James Woodburn's (1982a) distinction between "delayed return" and "immediate return" foraging societies is helpful. It is a distinction based on the systemic complexity of social structure. Delayed-return systems have more elaborate kinship structures such as lineages and clans, while immediate-return systems are based on simpler family structures. Immediate-return systems have economies in which people obtain an immediate yield for their labor and do not save food or property for later use. The San Ildefonso Agta in their Forager Phase were an immediate-return system. (See chapter 1.) Even in the Peasant Phase, most of them still show little indication of switching to a delayed-return economy. This discussion will be limited to this type of foraging society.

CRITERIA FOR SELECTION OF COMPARATIVE STUDIES

The selection of comparative cases from the forager literature poses a dilemma. Robert Kelly (1995:205–9) has clearly stated the problem:

> Any attempt to understand hunter-gatherer demography must confront the scanty data at our disposal. Given the importance often ascribed to population as a prime mover of cultural evolution, it is surprising that we have few accurate data on hunter-gatherer demography. Although we can compile a fairly impressive list. . . . The accuracy of many of these data is unknown for a variety of reasons. . . . All of these factors mean that we should regard hunter-gatherer [demographic] data with a healthy amount of skepticism.

Kelly's list exemplifies making surveys of the literature and extracting demographic averages based on them. These surveys are the "wide" approach to the problem. Kelly expresses his dissatisfaction with his own list. Constantly haunted by the same skepticism, some of which was expressed in chapter 5, we have formulated the following six methodological criteria in searching for comparable studies of hunter-gatherer populations that could be used in this chapter. The use of these criteria results in a "narrow" approach to the problem.

Criterion 1: Were the researchers able to acquire valid data? This is the most problematical aspect of this type of research. Anthropological populations typically have moved many times (and not always as a group) and are illiterate (and thus have no records). They may have different types of counting systems and are thus unaccustomed to extending quantification to as many things and events, unlike those who have been socialized in technological societies. It is extremely difficult and frequently impossible to obtain answers to questions of when, how many, and how often particular demographic events occurred. Long-term residence or contact by a person with ethnographic sensitivity is usually required to overcome these problems. This distinction will be of prime importance in the selection of cases for examination here.

Criterion 2: Were demographic and population data the main objective of the research? Given the difficulties of the research situation and the time and effort required to overcome them, demographic and population data usually must be the main focus of the effort. If not, other concerns take precedence and the demographic work is usually incomplete. This type of situation has been an important reason for the lack of reliable demographic data about anthropological populations.

Criterion 3: Was the demographic work comprehensive so that data were obtained on all the components of population change, or even complete with regard to one of these variables: fertility, mortality, and migration? This question addresses the analytical value of the data collected. As a result of the constraints discussed in the first two criteria, frequently the data collected on anthropological populations are disjointed and incomplete. In such cases the analytical power of the demographic equations cannot be used.

Criterion 4: Are the data longitudinal? The analysis of population dynamics requires knowledge of what has happened to a group over time. Again, the constraints of the field situation militate against acquiring longitudinal data. It is difficult to define the length of time required because it depends on the historical circumstances of the group and the questions being asked. Sometimes the assumption of stability is made so that a synchronic profile is presumed to have longitudinal validity. Unless a reasonable basis for making this assumption can be shown, this approach is highly questionable.

Criterion 5: Have the data been subjected to a critical process to probe their validity and to check that the above difficulties have been sufficiently surmounted? Chapter 5 has discussed this problem. Some type of logical system needs to be applied to the data to point out weaknesses and lacunae which are always present. The critical process needs to be explained. Much of the literature on the population dynamics of anthropological populations is lacking in any mention of it. This omission gives the impression that the data were simply gathered and presented without any verification.

Criterion 6: Have the demographic data been integrated with ethnographic and/or biological data so that there is some understanding of the cultural and biological processes behind the quantitative levels? This is required for analysis of population dynamics.

THE COMPARATIVE STUDIES

The goal was to select cases of immediate-return foragers that appeared to be methodologically sound. Such cases would reduce the skepticism exemplified in Kelly's statement quoted in the previous section. An examination of them and comparison with the results of this Agta study yield a demographic profile of fertility and mortality for such groups. This narrower approach with its emphasis on methodological soundness may have more analytical value than the wider approach with its emphasis on averages taken from broad surveys of the literature.

The six criteria described in the previous section were kept in mind in reviewing the literature. They are ideals never fully attained by any study, including this one. But they were approximated to various degrees by the demographic studies of these seven immediate-return foraging groups: the Ache, the Aka, the !Kung, the Hadza, the Batak, the Cagayan Agta, and the Casiguran Agta. Table 12.1 classifies these groups according to historical periods, as was done for the San Ildefonso Agta and the Acculturating population in chapters 6 and 11. Each column of the table contains methodological and demographic information for a group during a historical period. The selection of these groups was especially guided by the length of time the researcher spent with or remained in contact with the group studied, as stated in the first line for each group (row 1 of Table 12.1) and carries the label "Field Years." Table 12.1 compresses much information. The reader should read the notes at the bottom of that table to understand the abbreviations for the indices at the beginning of the rows.

The selection of these seven cases has some shortcomings. The number of cases is small. The question of how representative these cases are of all immediate-return foraging groups is not addressed. All the cases come from tropical or subtropical zones. Table 12.1 does not claim to be the result of a comprehensive search of all the literature, using the methodological criteria. Some important cases may have been omitted. Therefore, the resulting profile (Table 12.1) should be seen as an initial effort rather than a definitive formulation. In spite of these shortcomings, a comparison of these cases helps one to see the results of this study of the San Ildefonso Agta against a broader background.

THE FORAGER PHASE

The Studies

The six hunter-gatherer groups included in the Forager Phase column of Table 12.1 are the San Ildefonso Agta, the Cagayan Agta, the Ache of Paraguay, the Aka Pygmies of the Central African Republic, the !Kung Bushmen of southern Africa, and the Hadza in Tanzania. A question arises about the methodological comparability of the !Kung and Hadza studies with the others. The two Agta, the Ache, and the Aka studies were the result of field investigations of the demographic variables. The !Kung and Hadza studies were based on field studies of a few key variables to determine the selection of mathematical models that, in turn, projected the levels of the other variables.

There is a question about the applicability of these models to foragers. Usually the Coale-Demeny models are used, but as their formulators (Coale and Demeny 1966:29) stated, "There is no strong reason for supposing that the age patterns of mortality exhibited in these four families cover anything like the full range of variability in age patterns in populations under different circumstances." Therefore, when the models are used, it is difficult to assess the validity of the results.

In addition the models produce vital rates that "should be expected to characterize not a particular year but rather a much longer period, perhaps as long as a century. The stable population model, then, should describe the population on the average over many points of time, but not at every point" (Howell 1976:144). Therefore, the models average away the cultural and epidemiological history of a group and leave many questions unanswered.

The Agta and Ache studies are comprehensive studies of foragers and include field investigation of the demographic variables (although the Ache had no migration). They will be given greater weight here. The Cagayan Agta and Aka studies are surveys only of fertility.

Fertility

The Total Fertility Rate

In row 8 of Table 12.1 the total fertility rate (TFR) is shown for several populations. The rates in the Forager Phase range from 6.2 to 8.1 live births per woman except for the !Kung case. All these TFRs are based on field studies. A TFR of 8.0 was also found for the foraging Warao of the Orinoco Delta by Early and Heinen (Early 1985:394). Bailey and Aunger (1995:201) have found among the Efe, a Pygmy group in Zaire, a low TFR of 2.56, caused by sexually transmitted diseases that lead to sterility. These cases indicate that foragers have a high fertility level, unless there is a pathology disrupting the reproductive processes. The low !Kung TFR of 4.7 is partially influenced by the age at menarche. (*High* and *low* are used here to characterize these TFRs. However, there is no consensus about the quantitative equivalents for the qualitative adjectives *high, moderate,* or *low* with regard to any of the demographic variables.)

Table 12.1. Survey of Demographic Indices for Some Immediate-Return Foragers*

| | | Forager | | | | Phases | |
		1 Agta	2 Agta	3 Ache	4 Aka	5 !Kung	6 Hadza
1.	Field years	62–94	80–82	78–93	74–84	63–73	66–67
2.	Years of pop.	50–64	—	50–70	—	—	67
3.	# Years	15	—	21	—	—	1
4.	Person-years	2566	—	—	—	—	—
5.	Pop. begin	177	—	241	—	—	566
6.	Pop. end	221	—	547	—	—	566
Fertility							
7.	CBR	58.0	—	—	—	—	48.7
8.	TFR	7.0	6.5	8.1	6.2	4.7	6.2
9.	# Births	149	74	587	60	291	—
10.	Age menarche	—	17.1	15.3	—	16.6	—
11.	1st marriage	19.3	—	15.2	—	—	—
12.	1st birth	20.4	20.1	19.5	—	18.8	—
13.	Last birth	—	—	42.1	37.5	34.4	—
14.	Age menopause	—	43.9	—	—	—	—
15.	Birth interv.	2.8	2.4	3.1	3.1	4.1	—
16.	BI no mort.	3.0	2.9	—	—	—	—
17.	BI mort.	2.3	—	—	—	—	—
Age specific							
18.	15	121	—	159	—	115	—
19.	20	265	—	275	—	242	—
20.	25	370	—	298	—	203	—
21.	30	245	—	318	—	152	—
22.	5	208	—	279	—	119	—
23.	40	193	—	288	—	44	—
Mortality							
24.	e_0	24.3	—	36.9	—	34.6	31.5
25.	CDR	45.6	—	26.5	—	28.9	13.9
26.	# Deaths	117	—	352	—	—	—
q Probability							
27.	0	169	—	116	—	213	—
28.	1	128	—	139	—	146	—
29.	5	48	—	97	—	94	—
30.	10	56	—	102	—	18	—
31.	20	97	—	116	—	9	—
32.	30	344	—	114	—	117	—
33.	40	290	—	148	—	154	—
34.	50	533	—	280	—	131	—
35.	60	506	—	388	—	502	—
36.	70+	1,000	—	1,000	—	1,000	—
37.	Natural Inc	12.5	—	—	—	—	15.2
38.	Total Inc	16.7	—	—	—	—	—

*See following pages for explanations of columns and rows.

						Reservation	Acculturating from
	Transition			Peasant			
7	8	9	10	11	12	13	14
Agta	Ache	Hadza	Agta	Agta	Ache	Agta	Batak
62–94	78–93	77,85	62–94	62–86	77–93	62–86	66–81
65–79	71–77	67–85	80–94	77–84	78–93	52–94	72–80
15	7	18	14	7	16	42	8
2873	852	—	3087	—	5721	1244	1497
221	547	566	190	618	369	4	130
190	369	719	231	609	537	101	208
47.7	—	46.7	54.7	43.0	50.9	78.0	63.5
6.5	—	6.2	7.6	6.3	8.5	9.2	—
137	—	—	169	184	291	97	95
—	—	—	—	—	14	—	—
19.3	—	—	19.3	—	—	—	—
20.4	—	—	20.4	—	17.7	—	—
—	—	—	—	—	38.5	—	—
—	—	—	—	—	—	—	—
2.8	—	—	2.8	—	2.8	—	—
3.0	—	—	3.0	—	—	—	—
2.3	—	—	2.3	—	—	—	—
107	—	—	161	—	258	337	—
350	—	—	283	—	333	380	—
315	—	—	341	—	341	270	—
269	—	—	320	—	333	409	—
137	—	—	294	—	265	301	—
115	—	—	130	—	176	148	—
29.2	—	31.5	22.2	21.5	48.1	36.9	—
34.8	119.1	13.9	47.6	45.0	20.8	33.6	24.0
100	338	—	147	—	113	42	36
241	—	—	266	342	143	227	—
227	—	—	342	—	108	181	—
55	—	—	99	---	28	81	—
62	—	—	47	—	37	103	—
78	—	—	81	—	41	35	—
187	—	—	231	—	28	157	—
155	—	—	311	—	60	—	—
246	—	—	443	—	75	—	—
852	—	—	641	—	351	—	—
1,000	—	—	1,000	—	1,000	1,000	—
12.9	—	15.2	7.1	-2.0	30.1	44.2	39.5
-10.8	—	—	13.3	—	—	80.4	52.1

* Explanation of Columns and Rows in Table 12.1

Column 1: The San Ildefonso Agta during the Forager Phase. Because of the difficulty of obtaining exact dates, the indices for the lengths of the birth intervals (rows 16 and 17), ages at first marriage (row 11), and first birth (row 12) were drawn from all of the phases and are repeated in each phase, on the assumption that there has been little change in the fertility structure.

Column 2: Cagayan Agta (fertility data from Goodman et al. 1985:171–75).

Column 3: Ache hunter-gatherers in Paraguay from 1950 to 1970. Fertility data from Hill and Hurtado 1996:224, 230, 254, 261–62; most of the fertility cases are from the 21-year period from 1950 to the end of 1970 (Hill and Hurtado: personal communication). Mortality data from Hill and Hurtado 1996:171–73, 196–99; most of the mortality cases are from the 41-year period from 1930 to 1970 (Hill and Hurtado, personal communication). This table uses the mortality span for the indices: "Years of Pop.," "Pop. Begin," and "Pop. End" (rows 2, 5, and 6). The crude death rate (row 25) is the life table crude rate. The life table values given here are slightly different from those in the source, because the sexes have been combined and the figures recalculated to fit the age classification used in this table.

Column 4: Aka Pygmies in the Central African Republic. Data from Hewlett 1988:267 and Hewlett et al. 1988.

Column 5: !Kung Bushmen in Botswana and Namibia. Data from Howell 1979: 124, 128, 130, 134, 178, as well as data collected by the Harvard Kalahari research team over more than 10 years of fieldwork. Ages 40–44 and 45–49 are combined here. Lee (1979a:322) found slightly shorter birth intervals in his study. Mortality data from Howell 1979:88; these rates are slightly different from those in the source because of our recalculation.

Column 6: Hadza in Tanzania. Data from Blurton Jones et al. 1992:173 and Dyson 1977. A significant portion of the Hadza study is based on Woodburn's unpublished data. Woodburn has worked among or remained in contact with the Hadza for most of his professional life.

Column 7: San Ildefonso Agta during the Transition Phase.

Column 8: Ache during their transition stage. Mortality data from Hill and Hurtado 1996:175–76 for number of deaths. The denominators are annual population values obtained from Hill and Hurtado (personal communication in 1996).

Column 9: Hadza during their transition stage. Data from Blurton Jones et al. 1992:173 and Dyson 1977; unpublished data from Woodburn; and a 1977 census by Lars Smith, who spent four years among the Hadza.

Column 10: San Ildefonso Agta during the Peasant Phase.

Column 11: The Casiguran Agta from 1977 to 1984. Data from Headland 1989.

Column 12: Ache during their reservation period from 1977 to 1993. Data from Hill and Hurtado 1996:177, 201–3, 257; 262–63. For the crude rates (rows 7 and 25), the denominator was calculated from annual populations provided by Hill and Hurtado. Life table values are slightly different from the source because of our recalculation.

Column 13: The Acculturating population on the San Ildefonso Peninsula. Data from chapter 11 in this book.

Column 14: The Acculturating Batak Negrito population on Palawan Island, Philippines. Data based on Eder 1987 and Eder personal communication in 1996. Crude rates were calculated from a revision of the information in Table 11 in Eder (1987:110) so that it was comparable with the definition of the San Ildefonso acculturating population as defined in chapter 11 of this book. Eder furnished the researchers with the required information to make the recalculation possible.

Row 1: Field years—gives an approximation of the years the researcher was either in the field or in contact with the field situation. These figures should be read understanding "19–" before each (i.e., "62–94" means "from 1962 through 1994"). This row gives some sense of the researcher's familiarity with the culture indicated in that column. In many cases these are estimates. The authors and/or the sources should be checked for nuances and exact details.

Row 2: Years of pop.—gives the years covered by the demographic indices for the study population. This column is not applicable to all the studies.

Row 3: # Years—gives the number of years included in row 2. It is an index to indicate longitudinal studies. The Batak and Hadza study only attempted measurement at the beginning and end of the time periods. The San Ildefonso Agta and Ache studies are longitudinal in the full sense.

Row 4: Person-years—the number of person-years included in number of years shown in row 3. With collapsed data, this column gives a sense of the N used in the calculations.

Row 5: Pop. begin—the size of the study population at the beginning of the period.

Row 6: Pop. end—the size of the study population at the end of the period.

Row 7: CBR—crude birth rate.

Row 8: TFR—total fertility rate.

Row 9: # Births—number of births on which the above two indices (rows 7 and 8) are based.

Row 10: Age menarche—average age at menarche.

Row 11: 1st marriage—average or median age at first marriage.

Row 12: 1st birth—average or median age of the mother at the birth of her first child.

Row 13: Last birth—average age of the mother at the birth of her last child. In the literature there is ambiguity at times as to whether this means mothers who survive to menopause or all mothers, regardless of age at death. The average age for San Ildefonso Agta mothers throughout the 44-year period (age 39) was calculated from 46 mothers who survived to menopause (chapter 7).

Row 14: Average age of women at menopause.

Row 15: Birth interv.—average length in years of all measured birth intervals.

Row 16: BI No mort.—average length in years of birth intervals in which the initial birth during that interval survived the first 2.25 years of life; i.e., there was no infant or child mortality during the interval.

Row 17: BI mort.—average length in years of birth intervals in which the initial birth during that interval died in the first month of life.

Rows 18–23: Age Specific—age-specific fertility rates per 1,000 population.

Row 24: e_0—the life table function e at birth, the average life expectancy at birth in this population. Usually this and all life table functions are calculated for each sex. They have been combined here because of the small numbers.

Row 25: CDR—crude death rate.

Row 26: # Deaths—confer # of births above.

Row 27: The probability of dying during the first year of life.

Row 28. The probability of dying between age 1 and age 4.

Row 37: Natural inc.—natural increase.

Row 38: Total inc.—total increase equals the natural increase plus net migration.

The surveys of the TFR of all types of foragers by Campbell and Wood (1988), Bentley et al. (1993a), and the revision by Bentley et al. (1993b) found averages of 5.7, 5.6, and 5.4 respectively. The work of Bentley et al. is an extensive survey of the literature, but it does not escape the skepticism noted by Kelly. It includes in the forager category groups such as the Yanomami, who are primarily swidden agriculturalists, although foraging is also important in their subsistence pattern. Here we have preferred to restrict the consideration to immediate-return foragers, all of whom may do occasional horticulture. The lists of Bentley et al. (1993a, 1993b) also include cases that do not appear to adequately approximate the six criteria previously discussed.

Average Length of Birth Interval

The heading "Birth Interv." for row 15 in Table 12.1 includes all the types of birth intervals (BI) described in chapter 7 and Table 7.2. All the values of BIs in Table 12.1 are based on field studies. They yield an approximate three-year length for all types of intervals except the !Kung study. The Agta intervals are less than three years. However, the Agta studies are the only ones that distinguish the types of BIs. These distinctions help to explain the lower averages.

A BI in which the initial birth of the interval survives to 2.25 years is listed in row 16 of Table 12.1 as "BI No Mort.," that is, a birth interval without an intervening infant or child mortality. In chapter 7 and Table 7.2 this was called a regular interval or an interval with no early mortality. The Agta intervals of this type approximate the average length for all types of intervals of the other studies.

A BI in which the initial birth of the interval dies in the first month of life is labeled "BI Mort." in row 17 of Table 12.1, that is, a birth interval containing a neonatal mortality. Since a mortality at this age means the mother ceases nursing earlier than she would in a regular interval, the contraceptive effect of lactation ceases sooner, which often shortens the length of the BI Mort. The infant mortality rate of the San Ildefonso Agta is higher than that of the other groups. (See q_0 in row 27 under "Mortality" in Table 12.1.) Therefore, Agta BIs with infant mortality comprise a higher percentage of the distribution of all types of intervals (Birth Interv.), than do the percentages in the other groups. Because of this higher percentage, the Agta average for all types of intervals is lower than the same averages for the other groups.

The average BI of the !Kung is about four years. Their longer birth intervals have been attributed to the multiple effects of very frequent nursing and other biological factors (Konner and Worthman 1980:790; Konner and Shostak 1987:17). The hypothesis that the longer !Kung BI may be influenced by sexually transmitted diseases also appears a possibility (Harpending and Draper 1990:129).

Ages of the Reproductive Cycle

One of the determinants of fertility is the length of the female reproductive cycle. The age of menarche marks the beginning, and the age of menopause brings it to a close. One reason the Ache have the highest TFR is their longer reproductive period. Ache girls reach menarche almost two years before Agta girls and have their first child about one year earlier than Agta women. Variation in the age at menarche may be an important factor in the variation of TFRs among foragers. The Ache also have the highest age of last birth. The relatively early age of last birth for the !Kung may be a clue to the reason for their low fertility.

Age-Specific Rates of Fertility

Figure 12.1 graphs the age-specific fertility rates for three of the groups shown in Table 12.1. It shows the age distribution of Ache fertility is quite different from that of the Agta. The Ache have a higher fertility rate in the 15–19 age group than the Agta, reflecting their earlier age at menarche and first marriage. The usual expectation about the shape of the female fertility curve in noncontraceptive populations is a sharp rise during the 20s, to a peak in the 20–24 age category or sometimes in the 25–29 group. This peaking is followed by a noticeable decline through the 30s and ends in the early 40s. The Ache curve is distinctive because it is relatively flat with little variation from age 20 to the early 40s. The small peak comes in the age 30–34 range. As Hill and Hurtado (1996:253) note, the question is whether the source of this distinctiveness is estimation error in determining the birth dates of Ache mothers and their infants. If estimation error is minimal, the curve appears to represent a newly found age pattern distinctive of forager fertility. A comparison with the Agta and !Kung curves suggests that there are problems of age estimation in the Ache data. But the Agta and !Kung curves also contain some estimation error. Chapter 4 discussed the problem for the Agta data. The method used by Howell to assign ages has been questioned by Harpending and Wansnider (1982:37). More research is needed.

Summary of Fertility

The analysis shows that these immediate-return foragers have a total fertility rate of between six and eight live births per woman who lives to age 45, and a crude death rate under most conditions in the 50s. The average birth interval is about three years. This interval appears to be maintained primarily by lactation and its contraceptive effect. The Agta do not practice abortion or infanticide, and the Headlands came across only one case of venereal disease. The Ache have a low infanticide rate of 1.2 percent of all births, and it is also low among the !Kung, but in neither case is it done for the purpose of controlling the size of the population. There is no evidence of a systemic forager requirement for long birth intervals because of the necessity to transport children. This type of fertility profile, except the Ache age distribution, is not unique to foragers, but is also found in other types of preindustrial populations. The data indicate that the former axiom of low forager fertility is unfounded. The axiom appears to have been based on methodologically weak studies or speculation about effective fertility as a result of significant infanticide under the imperative of mobility, or stud-

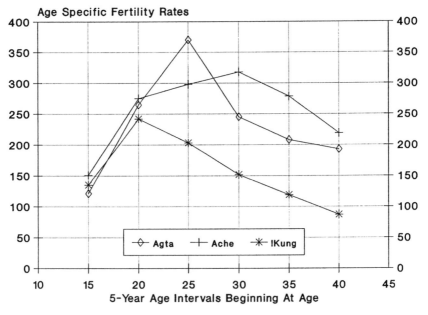

Figure 12.1. Age-specific fertility rates of Agta, Ache, and !Kung foragers.

ies of groups in which there was a biological pathology that hindered the reproductive system. The !Kung case, an exception for immediate-return foragers, is not completely unexpected because of the !Kung's location among many groups in the African low-fertility belt with a high incidence of sterility from sexually transmitted disease. Whatever may be the reason, it does not appear to be a result of their being foragers.

Levels of Mortality

Table 12.1 shows that the Ache have a lower level of mortality than the Agta. There is 12.6 years difference between the life expectancies at birth (e_0), while the Ache crude death rate is almost half that of the Agta (row 25). Figure 12.2 graphs the probability rates of dying by age (q) for both the Agta and Ache populations and for the !Kung. It shows that the higher levels of mortality for Agta infants and those over age 29 are responsible for the greater overall Agta mortality. However, there is some question about the quality of the age estimates used for the Ache life table, as seen earlier in the discussion of fertility.

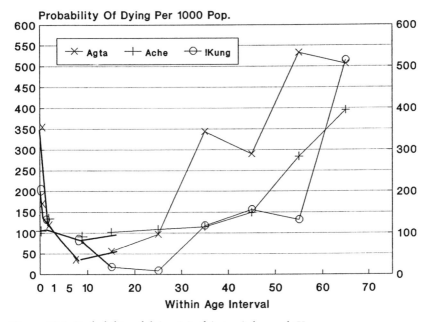

Figure 12.2. Probability of dying (q) of Agta, Ache, and !Kung.

Causes of Mortality

Are the differential levels of mortality of these three populations generated by similar causes of death with differing severity or by different types of mortality structures? The Ache case appears to have a distinctive structure.

The Ache and Homicidal Violence

The Ache causes of death are: violence (55 percent), illness caused by infectious diseases and parasites (23.5 percent), accidents (13.1 percent), and congenital/degenerative disease (7.6 percent) (Hill and Hurtado 1996:171–72). The role of violence leading to homicide in Ache mortality makes it a distinctively different structure from that of the San Ildefonso Agta, who have a moderate homicide rate of 129 per 100,000 population (see chapter 8 on mortality). About 40 percent of the Ache deaths by violence were by other Ache, and 60 percent were due to outside warfare with Paraguayan peasants. The Ache were under constant threat by kidnappers who wanted to use them as slave-workers. The majority of the deaths at the hands of peasants were children who had been kidnapped and were presumed to have died. Outside violence has been important in causing depopulation among small foraging groups in South America over the last 150 years (see Bodley 1990:38–41), but Hill and Hurtado are the first to provide demographic data on the contribution of this violence to the overall mortality of an Amerindian foraging population.

If the mortality caused by outside violence is removed from the Ache profile, does the percentage distribution of the remaining deaths appear similar to the Agta distribution? Hill and Hurtado's (1996:153, 155, 160) description mentions morbidity from diarrhea, respiratory infections, and parasites. They stress the problem of multiple attacks, which lower the resistance of the host to disease. These findings seem to indicate the same type of pathology as the San Ildefonso Agta.

With the Ache mortality from outside violence removed, the percentage distribution of the Simplified Table within Table 12.2 shows that the infectious and parasitic diseases still account for only 36 percent of the Ache mortality, which is far lower than the assumed San Ildefonso Agta level of 86 percent. Internal violence still accounts for 33 percent of all Ache deaths in the Simplified Table. Sixty percent of the mortality from internal violence is against young children three years of age or younger. If these children had died not by violence but from infectious and parasitic diseases, the Ache percentage of infectious and parasitic disease would reach 56 percent, still considerably less than the Agta level. Although mortality from violence is moderate among San Ildefonso Agta, Headland (1989:68) found

Table 12.2. Percentage Causes of Death of All Deaths for Four Cases of Immediate-Return Foragers

Cause	Original Tables				Simplified Table			
	Agta %	Ache %	Aka %	!Kung %	Agta %	Ache %	Aka %	!Kung %
Infectious	14.0	23.8	62.9	79.5	85.7	35.8	84.3	79.5
Degenerative	—	6.8	—	9.1	—	10.2	—	9.1
Nutritional	4.1	—	—	—	4.1	—	—	—
Maternal	4.4	.8	2.1	—	4.4	1.2	2.1	—
Accident	2.7	13.1	5.4	—	2.7	19.7	5.4	—
Homicide	3.0	55.5	I*	11.4	3.0	33.1	I*	11.4
Not classifiable	—	—	8.5	—	—	—	8.5	—
Unknown	71.7	—	21.4	—	—	—	—	—
Total	100	100	100	100	100	100	100	100

* I means that the percentage in the previous line includes the category of this line.
San Ildefonso Agta—from Table 8.2. Simplified table assumes unknown causes are due to infectious diseases.
Ache—from Hill and Hurtado 1996:Table 5.2. Simplified table removes mortality due to external warfare and recalculates.
Aka—from Hewlett et al. 1988:54. The original table above has collapsed from Hewlett's table the following causes under infectious: infectious and parasitic, acute, diarrhea, respiratory, infantile and early childhood. The simplified table assumes the unknown causes are infectious diseases.
!Kung—from Howell 1979:69. The simplified table is the same. Howell has already included unknown causes under infectious and parasitic in her original table.

high levels of mortality from violence among other Agta groups on the mainland in Casiguran.

Infectious and Parasitic Disease among Ache and Agta

Why are infectious and parasitic diseases so much lower among the Ache than the San Ildefonso Agta? No definite answer can be given, but there may be some clues. Parasite loads may be lighter. Hill and Hurtado (1996:154) note that "Ache move their camps almost every day, minimizing the accumulation of waste products and parasite populations." The Agta move less frequently. Studies of the Isabela Agta and the Cagayan Agta cited in chapter 1 showed they moved their camps only 10 or 20 times a year; the San Ildefonso Agta move less often than that. This difference may be an important factor because it suggests a distinction between "nomadic" foragers exemplified by the Ache and "lingering" foragers such as the Agta, who tend to remain longer at a campsite.

Another factor may be the differing climate of the two areas. Although the forest of the Ache in summer is tropical with high heat and humidity, there is also a winter season with occasional below-freezing temperatures. Dunn (1968:226) has speculated that tropical rain forests have more species of parasitic and infectious organisms than subtropical forest areas with frost. However, Dunn sees a lower prevalence of parasitic infections in tropical rain forests than freeze areas because the continual presence of parasites builds up the populace's immunity to these diseases. Therefore, the significance of the differing climate remains an open question.

Other Cases

So far it has been seen that both the levels and structures of Agta and Ache mortality are quite different. Which of these cases more closely resembles the Aka and !Kung cases? Table 12.1 shows the !Kung and Hadza studies more closely approximate the Ache life expectancy, although the difficulty of demographic comparisons using indices from models has been previously noted.

However, a comparison of the cause-of-death profiles in the simplified table within Table 12.2 shows that the Agta level of infectious disease more closely approximates the levels of the Aka and !Kung than does the Ache. The Aka case by Hewlett et al. (1988) is an excellent study of the problem of utilizing western medical categories for explaining the mortality of a culture that does not use these categories. In the Agta and Aka cases, the Simplified Table assumes the Unknown Cause category is composed mostly of deaths resulting from the dominant pathology indicated by the remainder of the data, that is, infectious disease. Howell (1979:69) had already made this assumption for her table about the !Kung.

Summary of Mortality

If the approximations in Table 12.2 based on the simplifying assumptions are accepted, the Aka and !Kung cases confirm the overwhelming importance of infectious and parasitic diseases, as indicated by the San Ildefonso Agta profile. The magnitude of the internal violence detracts from the significance of the other categories in the Ache case. The data confirm Dunn's (1968:225) observation: "Parasitic and infectious disease rates of prevalence and incidence are related to ecosystem diversity and complexity. Although many of these diseases contribute substantially to mortality, no simple, single generalization is possible for all hunter-gatherers."

As with fertility, the structure of mortality among these foragers does not appear unique, as parasitic and infectious disease are also important in peasant mortality (see Early 1982:96–130). The very high level of Agta mortality may be exceptional.

TRANSITIONAL FORAGERS

The Agta and Ache represent two very different types of transition from foraging societies. The Agta, who have been in close contact with agriculturists for centuries, have been through a gradual transition from a Forager Phase to a Peasant Phase, as described in chapters 2 and 3. Their demographic structure changed little in that 44-year period. Their lower mortality in this Transition Phase, owing to the medical intervention of the Headlands and New Tribes missionaries, was idiosyncratic, not systemic.

In contrast, transition for the Ache was abrupt. Their foraging period for at least 30 years prior to 1970 was a period of small-scale war with invading peasants who wanted to kidnap them and use them as laborers. The Ache themselves decided to leave the forest because of warfare and disease. As soon as one of the Ache groups emerged from the forest, they received assistance from local Paraguayans and missionaries. Initially the providers of this assistance were not prepared to handle the epidemics of infectious disease that struck the Ache, who lacked immunity. The crude death rate soared to 119 for the Ache during the first part of their transitional period in the early 1970s. Eighty-five percent of the adult deaths and 40 percent of the children's deaths were due to contact-related illnesses (Hill and Hurtado 1996:166). The deaths of a number of children were indirectly related to contact illnesses. They were buried alive with their deceased parents because there was no one to take care of them. Thirty-one percent of the precontact population perished during the Ache transitional period from 1971 to 1977 (Hill and Hurtado 1996:166–167).

We have classified the Hadza studies as covering a transitional period because of the references to some Hadza groups becoming sedentary, taking up farming, and engaging in wage labor, while other Hadza groups remained foragers. The second Hadza demographic study (Blurton Jones et al. 1992) found no differences in the demographic structure from the first one (Dyson 1977), when they were still more traditional. Lee's (1979a:401–31) study of the !Kung shows them also to be in transition from foraging to farming, cattle raising, and wage labor. His data (1979:322; 1984) show that the length of the birth intervals with no mortality for transitional !Kung has fallen to two and a half years. This decline is because mothers have shortened the length of time they breast-feed their babies, the principal reason for which appears to be the abundance of milk at the cattle posts.

FORMER FORAGERS WHO HAVE BECOME PEASANTS

The San Ildefonso Agta fertility and mortality during their Peasant Phase (column 10 of Table 12.1) have been described in chapters 7 and 8. The second Agta study (column 11) is an earlier demographic survey of the Casiguran Agta by Headland (1989). Casiguran Agta here is a language designation. It includes any Agta who speak Casiguran Agta as their first language, regardless of residence. Most of them live in the northern part of Aurora Province, including the San Ildefonso Peninsula. Therefore, there is some overlap with the people included in this study. The time frame of the Casiguran Agta survey is 1977–84, which straddles the Transition and Peasant Phases of the San Ildefonso study in this book.

The time frame of the Casiguran Agta survey has been classified as Peasant, because social change came earlier to the mainland than to the peninsula. In the preliminary Casiguran Agta survey, crude rates of fertility and mortality are slightly lower than those of the San Ildefonso Agta. The Casiguran study was conducted with the care typical of the better anthropological studies of foraging demography. However, the tighter methodological controls used by this later study of the San Ildefonso Agta yielded some missing children, which accounts for some of the difference. The seven-year span of the preliminary 1977–84 survey is too short a time period by which to judge a long-term trend. The negative natural increase of -0.2 percent per year that Headland reported for the Casiguran Agta for that seven-year period can also be found for numerous seven-year segments of this peninsula study (see Figure 6.1). Overall, the Headlands' 1977–84 Casiguran Agta preliminary survey confirms the high levels of fertility and mortality of the San Ildefonso Agta during the 1980–94 Peasant Phase.

Although not listed in Table 12.1, a substantial number of !Kung have also become sedentary, by the late 1970s engaged in wage labor, stock raising, craft sales, and horticulture. Harpending (Harpending and Wansnider 1982:39) has surveyed the fertility of both foraging and sedentary !Kung groups and found them to be similar, which agrees with the Agta finding. His results are not included in Table 12.1 because of the use of a noncomparable measure discussed in chapter 6. Harpending surveyed !Kung working in farming areas, not cattle posts, and this may be the reason his result appears to differ from Lee's finding of decreasing length of birth intervals.

FORAGERS ON RESERVATIONS

With the assistance of the government and especially of missionaries, the Ache were settled on six reservations and protected from intrusion by outsiders. In 1988 they received legal title to the reservations, which are on the perimeter of their traditional foraging areas to which they periodically return. They also engage in horticulture and wage labor and receive competent medical services. Governmental presence has brought an end to both external and internal violence. The Ache are focused almost exclusively on their own communities and show little concern with their status in Paraguayan society. Few outgroup marriages have taken place. Hill and Hurtado (1996:74) have labeled them a "settlement/trekking population." It appears that as long as they can keep their present independent economic base and maintain their separation from rural Paraguayan society, they may be able to escape incorporation into the national structure as landless peasant laborers.

The Ache situation stands in sharp contrast to that of the San Ildefonso Agta and is reflected in the mortality levels of the two groups in Table 12.1. Ache life expectancy has increased by 11 years to a level that is more than twice that of the Agta. The Ache still suffer from infectious diseases, but they receive western medical treatment. The relationship between their infant and child mortality rates is the reverse of the Agta, which indicates the absence of the synergism of malnutrition and infectious disease. The Ache fertility rate continues to be high and has slightly increased. Therefore, their population is increasing at a high annual rate of 3 percent, while the Agta are struggling to maintain themselves with a rate of 0.2 percent. These contrasts between the two groups indicate the importance of providing basic public services to tribal and peasant populations in frontier areas.

ACCULTURATING POPULATIONS

The Batak case is classified with the San Ildefonso Acculturating population described in the previous chapter. The Batak are Negritos on the island of Palawan in the Philippines who have extensively intermarried with Tagbanuas (a neighboring non-Negrito tribal people) and with lowland Filipino colonists. In the 1980s the Batak still engaged in some foraging, but it was one of many kinds of economic activity, including agriculture and wage labor (Eder 1987:52). Eder has furnished us with unpublished data that allow us to revise his Table 11 (Eder 1987:110) so that the definitions are the same as those used in this study. The Batak population as he defines it can be subdivided into Batak married to each other with their children (comparable to the San Ildefonso Agta population) and Batak married to Tagbanuas or Filipinos with their children (the Batak Acculturating population).

Table 12.1 shows the crude rates for the Batak Acculturating population. Fertility (following row 7) is very high in both Acculturating populations, with some of the difference probably accounted for by differences in the age-sex structures. There is continuing high in-migration for marriage in both groups, as can be assumed from the high level of net migration for the Batak. Eder's data (personal communication) show that 75 percent of the Batak married to Tagbanuas or Filipinos are females (hyper-gyny) and that access to land is part of the attraction. It appears that the Batak are headed for assimilation in the same way as the San Ildefonso Agta. Eder in his aptly titled book *On the Road to Tribal Extinction* concentrates on the declining number of Batak. But as Eder makes clear, this decline is not primarily due to mortality but to the declining number of Batak marrying other Batak. Thus the Batak case confirms the demographic profile of the San Ildefonso Acculturating population of this study.

SUMMARY

This final comparative chapter has attempted to put the results of the San Ildefonso Agta study into the wider perspective of some other studies of forager population dynamics selected by methodological criteria. It has provided a demographic profile of these foraging groups, which can serve as a basis of discussion and further research. The authors believe the research has contributed to a better understanding of forager fertility, mortality, and the process of change away from the foraging stage. It has also highlighted the importance of certain methodological and policy propositions, some of which have been summarized here.

Methodology

This study and the Ache study are the only ones that have attempted to reconstruct all of the demographic variables of a foraging population during their traditional foraging periods. Both studies relied heavily on the reconstruction of fertility histories and subjected them to a lengthy critical process. Both studies show the necessity of long-term involvement with the populations being studied.

Ex-Foragers as Peasants

The incorporation of the San Ildefonso Agta into the national structure did not involve any substantial changes in either fertility or mortality. The fertility structure appears to remain much the same as in the Forager Phase. This observation is contrary to a frequent statement that forager fertility rises with sedentism. It may in some cases, but not the Agta's. It is not clear if the statement assumes low fertility in the forager period or not. The Agta study may indicate a shift in the structure of mortality in this phase. The synergism of infectious disease and malnutrition, caused by the degradation of the rain forest by logging companies, becomes a major factor. In turn, this situation has resulted in the impoverished economic conditions of the Agta as landless agricultural workers. It has also contributed to the increase of infectious disease resulting from the increased population density of the peninsula as a result of lowlander in-migration.

Some Public Policy Implications

The results of this research may help public officials understand the need or consequences of certain public policies. This book is a window looking in on a world that is little known or even imagined by those immersed in the benefits of technological development. Although foragers are a minuscule percentage of the world's population, the peasantry, of which they are becoming a part, are not. A case could be made that they are the world's majority.

An important policy lesson is the necessity of basic health programs in these remote areas to enable their populations to combat the high levels of mortality. The Agta case has shown how a small medical program impacted Agta mortality during the Transition Phase. This example is reinforced by the tremendous improvement in life expectancy among the Ache in their reservation period since 1977. In both cases, missionaries played an important role, working in conjunction with the national health services. Their on-the-scene presence appears to have been the key factor.

The story of the San Ildefonso Agta revealed here may be seen as a case history of the social consequences of the deforestation crisis, which is taking place on a global scale. While issues of global warming, rising ocean levels, and large-scale climatic changes are certainly important, the more immediate human problems may be overlooked (Sponsel, Headland, and Bailey 1996). The Agta help us to understand this dimension.

Most fundamental of all, the conditions elaborated in this research are a microcosm of the problems of war or peace for the global community. The Agta are a very small part of the global picture, but their problems at the approach of the millennium are the problems of many. Social justice demands a careful intellectual scrutiny of these problems and the political will to wrestle with them.

References Cited

Anderson, James N.

    1964   Land and society in a Pangasinan community. In *Social foundations of community development,* ed. Socorro Espiritu and Chester Hunt, 171–92. Manila: R. C. Garcia Publishing House.

Bailey, Robert C., and R. V. Aunger

    1995   Sexuality, infertility, and sexually transmitted disease among farmers and foragers in Central Africa. In *Sexual nature sexual culture,* ed. Paul Abramson and Steven Pinkerton, 195–222. Chicago: University of Chicago Press.

Barry, Capt. E. B.

    1901   Official report of Capt. E. B. Barry, USN. Typescript dated March 1901. Correspondence of Naval Secretary. Archived at National Archives and Records Service, Washington, D.C.

Bentley, Gillian R., Grazyna Jasienska, and Tony Goldberg

    1993a   The fertility of agricultural and nonagricultural traditional societies. *Population Studies.* 47:269–81.

    1993b   Is the fertility of agriculturists higher than that of nonagriculturalists? *Current Anthropology.* 14:778–86.

Birdsell, Joseph B.

    1968   Some predictions for the Pleistocene based on equilibrium systems among recent hunter-gatherers. In *Man the hunter,* ed. R. Lee and I. DeVore, 229–40. Chicago: Aldine Publishing Company.

Black, F.

    1975   Infectious diseases in primitive societies. *Science.* 187:515–18.

Blurton Jones, Nicholas G., Lars C. Smith, James F. O'Connell, Kristen Hawkes, and C. L. Kamuzora

    1992   Demography of the Hadza, an increasing and high density population of savanna foragers. *American Journal of Physical Anthropology.* 89:159–81.

Bodley, John H.
   1990   *Victims of progress.* 3rd edition. Mountain View, Calif.: Mayfield.
Burch, Ernest S., and L. Ellanna
   1994   *Key issues in hunter-gatherer research.* Oxford: Berg Publishers.
Campbell, K. L., and J. W. Wood
   1988   Fertility in traditional societies. In *Natural human fertility: Social and biological mechanisms,* ed. P. Diggory, S. Teper, and M. Potts, 39–69. London: Macmillan.
Carr-Saunders, A. M.
   1922   *The population problem: A study in human evolution.* London: Oxford University Press.
Casala, Tomas C.
   1960   *General census of the non-Christian tribes living in the municipality of Casiguran, Quezon.* Oct. 12, 8-page typescript. Casiguran: Report to the Commission of National Integration.
Census of the Philippines, various years
   n.d.   *Census of the population.* Manila: National Census and Statistics Office.
Clark, Constance
   1990   Trading networks of northeastern Cagayan Agta Negritos. M.A. thesis, University of Hawaii.
Coale, Ansley J., and Paul Demeny
   1966   *Regional model life tables and stable populations.* Princeton: Princeton University Press.
Constantino, Renato
   1975   *A history of the Philippines.* New York: Monthly Review Press.
De Bevoise, Ken
   1995   *Agents of apocalypse: Epidemic disease in the colonial Philippines.* Princeton: Princeton University Press.
Dumond, Don E.
   1975   The limitation of human population: A natural history. *Science.* 187:713–21.
Dunn, Frederick
   1968   Epidemiological factors: Health and disease in hunter-gatherers. In *Man the hunter,* ed. R. Lee and I. DeVore, 221–28, Chicago: Aldine Publishing Company.
Dyson T.
   1977   The demography of the Hadza. In *African historical demography: Proceedings of a seminar held in the Centre of African Studies,* 139–54. Edinburgh: University of Edinburgh.
Early, John D.
   1982   *The demographic structure and evolution of a peasant system: The Guatemalan population.* Boca Raton: University Press of Florida.

1985   Low forager fertility: Demographic characteristic or methodological artifact? *Human Biology.* 57:387–99.

Early, John D., and John F. Peters
1990   *The population dynamics of the Mucajai Yanomama.* San Diego, Calif.: Academic Press.

Eder, James F.
1987   *On the road to tribal extinction.* Berkeley: University of California Press.

Estioko-Griffin, Agnes, and P. Bion Griffin
1975   The Ebuked Agta of northeastern Luzon. *Philippine Quarterly of Culture and Society.* 3:237–44.

Forest Management Bureau
1989   *Philippine forestry statistics, 1989.* Philippines: Department of Environment and Natural Resources, Forest Management Bureau.

Goddard, George W.
1930   The unexplored Philippines from the air. *National Geographic.* 58:311–43.

Goodman, Madeline J., Agnes Estioko-Griffin, P. Bion Griffin, and John S. Grove
1985   Menarche, pregnancy, birth spacing and menopause among the Agta women foragers of Cagayan Province, Luzon, the Philippines. *Annals of Human Biology.* 12:169–77.

Gordon, John, John B. Wyon, and Werner Ascoli
1967   The second year death rate in less developed countries. *American Journal of Medical Science.* 254:357–80.

Griffin, P. Bion
1984a   Forager resource and land use in the humid tropics: The Agta of northeastern Luzon, the Philippines. In *Past and present in hunter gatherer studies,* ed. C. Schrire, 95–121. Orlando, Fla.: Academic Press.
1984b   Agta forager women of the Philippines. *Cultural Survival Quarterly.* 8:21–23.
1991   Philippine Agta forager-serfs: Commodities and exploitation. In *Cash, commoditisation and changing farmers,* ed. Nicolas Peterson and Toshio Matsuyama, 199–222. Osaka: National Museum of Ethnology, Senri Ethnological Studies 30.
1994   Becoming Filipino peasants: Agta forager gender role changes. In vol. 1 of *Hunters and gatherers in the modern context: Book of presented papers,* ed. Linda J. Ellanna, 234–38. Fairbanks: University of Alaska.

Griffin, P. Bion, Madeleine Goodman, Agnes Estioko-Griffin, and John S. Grove
1992   Agta women hunters: Subsistence, child care and reproduction. In *Man and his culture—A resurgence,* ed. Peter Bellwood, 173–99. New Delhi: Books and Books.

Griffin, P. Bion, and Thomas Headland
1994   The Agta of the Philippines. In *Traditional peoples today.* Vol. 5 of *The*

illustrated history of mankind, ed. Gorn Bruenhault, 71–73. New York: HarperCollins and American Museum of Natural History.

Harpending, Henry
1976   Regional variations in !Kung populations. In *Kalahari hunter-gatherers,* ed. R. Lee and I. DeVore, 152–65. Cambridge: Harvard University Press.

Harpending, Henry, and Patricia Draper
1990   Comment. *Current Anthropology.* 31:127–29.

Harpending, Henry, and LuAnn Wansnider
1982   Population structures of Ghanzi and Hgamiland !Kung. In *Ecology and population structures,* vol. 2 of *Current developments in anthropological genetics,* ed. Michael Crawford and James Mielke, 29–50. New York: Plenum Press.

Headland, Thomas N.
1978   Cultural ecology, ethnicity, and the Negritos of northeastern Luzon: A review article. *Asian Perspectives.* 21:127–39.

1985a   Imposed values and aid rejection. In *The Agta of northeastern Luzon: Recent studies,* ed. P. Bion Griffin and Agnes Estioko-Griffin, 102–18. Cebu City, Philippines: San Carlos Publications, University of San Carlos.

1985b   International economics and tribal subsistence: A microeconomic study of a Negrito hunter-gatherer society in the wake of the Philippine crisis of 1983. *Philippine Quarterly of Culture and Society.* 13:235–39.

1986   Why foragers do not become farmers: A historical study of a changing ecosystem and its effect on a Negrito hunter-gatherer group in the Philippines. Diss. no. 8622099. Ann Arbor: University Microfilms International.

1987a   Negrito religions: Negritos of the Philippine Islands. In vol. 10 of *The Encyclopedia of Religion,* ed. Mircea Eliade, 348–49. New York: Macmillan.

1987b   Kinship and social behavior among Agta Negrito hunter-gatherers. *Ethnology.* 26:261–80.

1988a   Ecosystemic change in a Philippine tropical rainforest and its effect on a Negrito foraging society. *Tropical Ecology.* 29:121–35.

1988b   Why foragers do not become farmers: The competitive exclusion principle and the persistence of the professional primitive. Paper presented at the 12th International Congress of Anthropological and Ethnological Sciences, Zagreb, Yugoslavia, July 24–31.

1989   Population decline in a Philippine Negrito hunter-gatherer society. *American Journal of Human Biology.* 1:59–72.

Headland, Thomas N., and P. Bion Griffin
1997   A bibliography of the Agta Negritos of eastern Luzon, Philippines. *SIL Electronic Working Papers* 1997–004. Dallas: Summer Institute of Linguistics. Online. URL: http://www.sil.org/wilewp/1997/004/silewp 1997-004.html. June 4, 1997.

Headland, Thomas N., and Janet D. Headland

1974a   *A Dumagat (Casiguran)—English dictionary.* Canberra: Australian National University.

1974b   Grammatical sketch of Casiguran Dumagat. *Pacific Linguistics.* A–43:1–54.

1997   Limitation of human rights, land exclusion and tribal extinction: The Agta Negritos of the Philippines. *Human Organization.* 56(1):79–90.

Headland, Thomas N., and Lawrence A. Reid

1989   Hunter-gatherers and their neighbors from prehistory to the present. *Current Anthropology.* 30:43–66.

1991   Holocene foragers and interethnic trade: A critique of the myth of isolated independent hunter-gatherers. In *Between bands and states: Interaction in small scale societies,* ed. Susan Gregg, 333–42. Carbondale: University of Southern Illinois.

Hewlett, Barry S.

1988   Sexual selection and paternal investment among Aka pygmies. In *Human reproductive behavior,* ed. Laura Betzig, Monique B. Mulder, and P. Turke, 263–76. Cambridge: Cambridge University Press.

Hewlett, Barry S., Jan M. H. Van De Koppel, and Luigi Luca Cavalli-Sforza

1988   Causes of death among Aka Pygmies of the Central African Republic. In *African Pygmies,* ed. Luigi L. Cavalli-Sforza, 45–63. San Diego, Calif.: Academic Press.

Hill, Kim, and A. Magdalena Hurtado

1996   *Ache life history.* Hawthorne, N.Y.: Aldine de Gruyter.

Hollnsteiner, Mary R.

1970   Reciprocity in the lowland Philippines. In *Four readings on Philippine values,* ed. F. Lynch and A. de Guzman II. 3rd ed. IPC Papers no. 2, 65–88. Quezon City: Ateneo de Manila Press.

Howell, Nancy

1976   The population of the Dobe !Kung. In *Kalahari hunter-gatherers,* ed. R. Lee and I. DeVore, 137–51. Cambridge: Harvard University Press.

1979   Demography of the Dobe !Kung. New York: Academic Press.

Ingold, Tim, D. Riches, and J. Woodburn, eds.

1988   *Hunters and gatherers* (2 vols.). New York: Berg Publishers.

Jones, Gregg R.

1989   *Red revolution.* Boulder, Colo.: Westview Press.

Kelly, Robert L.

1995   *The foraging spectrum.* Washington, D.C.: Smithsonian Institution Press.

Kent, Susan, ed.

1996   *Cultural diversity among twentieth century foragers.* Cambridge: Cambridge University Press.

Kerkvliet, Benedict J.

1977   *The Huk rebellion.* Berkeley: University of California Press.

Kessler, Richard J.
    1989   *Rebellion and repression in the Philippines.* New Haven: Yale University Press.

Knauft, B. M.
    1987   Reconsidering violence in simple human societies. *Current Anthropology.* 28:457–500.

Konner, M., and M. Shostak
    1987   Timing and management of birth among the !Kung: Biocultural interaction in reproductive adaptation. *Cultural Anthropology.* 2:11–28.

Konner, M., and C. Worthman
    1980   Nursing frequency, gonadal function and birth spacing among !Kung hunter-gatherers. *Science.* 207:788–91.

Krzywicki, Ludwig
    1934   *Primitive society and its vital statistics,* trans. H. E. Kennedy and A. Truszkowski. London: Macmillan.

Kummer, David M.
    1991   *Deforestation in the postwar Philippines,* Geography research paper no. 234. Chicago: University of Chicago Press.

Lee, Richard
    1979a   *The !Kung San: Men, women, and work in a foraging society.* New York: Cambridge University Press.
    1979b   Hunter-gatherers in process: The Kalahari Research Project, 1963–1976. In *Long-term field research in social anthropology,* ed. George Foster et al., 303–21. New York: Academic Press.
    1981   Is there a foraging mode of production? *Canadian Journal of Anthropology.* 2:13–19.
    1984   *The Dobe !Kung.* New York: Holt, Rinehart and Winston.
    1992   Art, science or politics? The crisis in hunter-gatherer studies. *American Anthropologist.* 94:31–54.

LeVine, Robert A.
    1977   Child rearing as cultural adaptation. In *Culture and infancy,* ed. P. H. Leiderman, S. R. Tulkin, and A. Rosenfeld, 15–27. New York: Academic Press.

Lynch, Frank
    1970   Social acceptance reconsidered. In *Four readings in Philippine values,* ed. F. Lynch and A. de Guzman II. 3rd ed. IPC Papers No. 2, 1–64. Quezon City: Ateneo de Manila Press.

Lynch, Owen J.
    1983   The Philippine indigenous law collection: An introduction and preliminary bibliography. In *Philippine Law Journal,* vol. 43, fourth quarter (December):457–534.

Masnick, George S., and S. H. Katz
   1976   Adaptive childbearing in a North Slope Eskimo community. *Human Bi-
           ology.* 48:37–58.
Meigs, Grace L.
   1917   *Maternal mortality.* Washington, D.C.: Government Printing Office.
           Reprinted in *Children's bureau studies,* 1974. New York: Arno Press.
NAMRIA-National Mapping and Resource Information Authority
   1991   *Report on forest cover survey and mapping of Aurora Province.* Manila:
           Department of Environment and Natural Resources, Aurora Integrated
           Area Development Project.
Parumog, Benedicto P.
   1982   Schedule for market values 1981–1982 for the Province of Aurora. (Bound
           typescript, 137 pp.)
Pennettii, V., L. Sgaramella-Zonta, and P. Astolfi
   1986   General health of the African Pygmies of the Central African Republic.
           In *African Pygmies,* ed. L. L. Cavalli-Sforza, 127–41. San Diego, Calif.:
           Academic Press.
Peterson, Jean Treloggen
   1978   *The ecology of social boundaries,* Illinois Studies in Anthropology, no.
           11. Urbana: University of Illinois Press.
Peterson, Nicholas
   1993   Demand sharing: Reciprocity and the pressure for generosity among for-
           agers. *American Anthropologist.* 95:860–74.
PCGS, Solano
   1986   Map #2508, Solano. Manila: Philippine Coast and Geodetic Survey.
Population Reference Bureau
   1995   *World population data sheet.* Washington, D.C.: Population Reference
           Bureau.
Rai, Navin K.
   1990   *Living in a lean-to: Philippine Negrito foragers in transition.* Museum of
           Anthropology, Anthropological Papers no. 80. Ann Arbor: University of
           Michigan.
Reid, Lawrence A.
   1987   The early switch hypothesis: Linguistic evidence for contact between
           Negritos and Austronesians. *Man and Culture in Oceania.* 3:41–59.
   1994   Unraveling the linguistic histories of Philippine Negritos. In *Language
           continuity and change in the Austronesian world,* ed. T. Dutton and D.
           Tyron, 443–75. Berlin: Monton de Gruyter.
Rosaldo, Renato
   1980   *Ilongot headhunting: A study in society and history.* Stanford: Stanford
           University Press.

Rose, Frederick G.
   1960   *Classification of kin, age structure and marriage amongst the Groote Eylandt Aborigines: A study in method and theory of Australian kinship.* New York: Pergamon Press.
Salomon, J. B., L. F. Mata, and J. Gordon
   1968   Malnutrition and the communicable diseases of childhood in rural Guatemala. *American Journal of Public Health.* 58:505–16.
Scanland, Patrick
   1976   The age computer: A simple device for improving age determination in censuses and surveys. *Public Health Reports.* 91:360–67.
Scrimshaw, N. S., C. B. Taylor, and J. E. Gordon
   1968   Interactions of nutrition and infection. Geneva: WHO Monograph Series no. 57.
Shapiro, Sam, E. R. Schlesinger, and R. E. L. Nesbitt
   1968   *Infant, perinatal, maternal and childhood mortality in the United States.* Cambridge: Harvard University Press.
Shryock, Henry S., and Jacob Siegel and Associates
   1973   *The Methods and Materials of Demography.* Second printing, revised. Washington, D.C.: U.S. Government Printing Office.
Soeda, Yukichi
   1985   Personal communication to Thomas N. Headland.
Spielmann, Katherine A., and James F. Eder
   1994   Hunters and farmers: Then and now. *Annual Review in Anthropology.* 23:303–23.
Sponsel, Leslie E., Thomas N. Headland, Robert C. Bailey, eds.
   1996   *Tropical deforestation: The human dimension.* New York: Columbia University Press.
Takamiya, Teiji
   1975   Rusonni Kiyu. Tokyo: Shioruma Shuppan Co.
Tobias, P. V.
   1964   Bushmen hunters-gatherers: A study in human ecology. In *Ecological studies in southern Africa,* ed. D. H. S. Davis, 69–86. The Hague: Mouton.
Truswell, A. Stewart, and John D. Hansen
   1976   Medical research among the !Kung. In *Kalahari hunter-gatherers,* ed. R. Lee and I. DeVore, 166–94. Cambridge: Harvard University Press.
Turnbull, Major Wilfrid
   1929   The Dumagats of north-east Luzon. *Philippine Magazine.* 26:131–33, 175–78, 208–9, 237–40.
   1930   Bringing a wild tribe under government control. *Philippine Magazine.* 26:782–83, 794–98; 27:31–32; 36–42, 90–91, 116–20.

United Nations
  1978   *Population of the Philippines.* Bangkok: United Nations, Economic and
          Social Commission for Asia and the Pacific.
  19—   *Demographic yearbooks.* New York: Department of Economic and So-
          cial Affairs, Statistical Office, United Nations.
Vanoverbergh, Morice
  1937   Some undescribed languages of Luzon. Nijmegen, Netherlands: Dekker
          van de Vegt N. V. *(Publications de la Commission d'Enquete
          Linguistique . . . 3.)*
  1937–38 Negritos of eastern Luzon. *Anthropos.* 32:905–28; 33:119–64.
Wiessner, Polly
  1982   Risk, reciprocity and social influences on !Kung San economics. In *Poli-
          tics and history in band societies,* ed. E. Leacock and R. Lee, 61–84.
          Cambridge: Cambridge University Press.
Williams, Ciceley, N. Baumslag, and D. Jelliffee
  1994   *Mother and child health.* New York: Oxford University Press.
Wolf, Eric
  1969   *Peasant wars of the twentieth century.* New York: Harper and Row.
Wood, James
  1987   Problems applying model fertility and mortality schedules to data from
          Papua New Guinea. In *The survey under difficult conditions: Demo-
          graphic data collection in Papua New Guinea,* ed. T. McDevitt, 371–
          97. New Haven: HRAF Press.
Woodburn, James
  1980   Hunters and gatherers today and reconstruction of the past. In *Soviet
          and western anthropology,* ed. Ernest Gellner, 95–117. New York: Co-
          lumbia University Press.
  1982a   Egalitarian societies. *Man.* 17:431–51.
  1982b   Social dimensions of death in four African hunting and gathering soci-
          eties. In *Death and the regeneration of life,* ed. M. Bloch and J. Parry,
          187–210. Cambridge: Cambridge University Press.
Worcester, Dean C.
  1912   Head-hunters of northern Luzon. *National Geographic.* 23:833–48

Index